IN THE COMPANY OF MUSHROOMS

ELIO SCHAECHTER

IN THE
COMPANY
OF $\mathbf{M}$USHROOMS

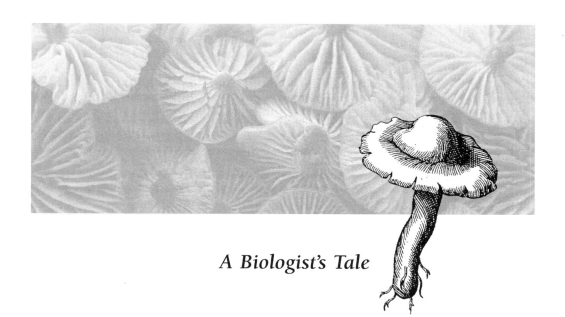

A Biologist's Tale

HARVARD UNIVERSITY PRESS · Cambridge, Massachusetts · London, England · 1997

Design by Marianne Perlak

Caution: This book is not intended as a recipe book or an identification guide. There are risks involved in consuming wild mushrooms. To minimize them, you must obtain positive identification of each specimen. Even with proper identification, the possibility exists that the consumer may be allergic to a mushroom, or that the mushroom may in some way be anomalous. The author has been conscientious in his efforts to alert the reader to potential hazards of consuming wild mushrooms, but the reader must accept full responsibility for deciding to consume any particular specimen. Descriptions of medicinal uses of mushrooms given in this book are for educational purposes only. The author is not recommending the use of mushrooms for self-medication. Always consult a physician about such use.

Library of Congress Cataloging-in-Publication Data
Schaechter, Moselio.
    In the company of mushrooms : a biologist's tale / Elio
Schaechter.
        p.   cm.
    Includes bibliographical references and index.
    ISBN 0-674-44554-6 (cloth : alk. paper)
    1. Mushrooms. 2. Mushrooms, Edible. I. Title.
 QK617.S32   1997
 579.6—dc21 96-29568

To Edith

CONTENTS

Acknowledgments *ix*

Prologue: The Hunt *xi*

MUSHROOMS AND OTHER FUNGI

1. Mushrooms in Our Midst *3*
2. The Life History of the Mushroom *23*
3. Umbrellas and Other Variations *41*

COLLECTING, IN SOLITUDE
AND IN GROUPS

4. A Foray in the Woods *57*
5. A Walk on the Lawn *85*
6. Mushroomers United *97*

CULINARY TALES

7. A Matter of Taste *123*
8. White Buttons, Shiitake, and Other Tamed Species *145*
9. Truffles, Stinkhorns, and Corn Smut *161*

A KINGDOM OF VERSATILE PARTS

10. Mushrooms, the Mind, and the Body *181*
11. Murder and More Mushroom Mayhem *201*
12. The Train Wrecker and Other Sturdy Mushrooms *220*
13. Insects as Fungus Gardeners *233*

Epilogue: The Biologist as Mushroom Hunter *247*

Resources *255*
Credits *273*
Index *275*

Color illustrations follow p. 144.

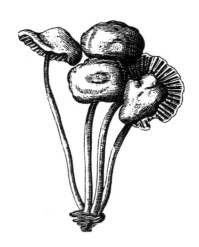

ACKNOWLEDGMENTS

I was aided in writing this book by the encouragement and help given by many friends who spent considerable time with the manuscript and provided incisive and insightful suggestions. For being so supportive and helping me both explicitly and implicitly, I wish to thank Donna Besecker, Sarah Boardman, Douglas Brown, Jean Cargill, Harry Davidow, Karen Davis, George Davis, Jack Fellman, Al Ferry, Joan Finger, Joanne Gilbert, Ilona Karmel, Geraldine Kaye, Jhanara Kriston, Milton Landowne, Larry Millman, Patty Morrison, Linda Napora, Lynn Payer, Donald Pfister, Tim Satterfield, Linc Sonenshein, Jack Walsh, Vivian Walworth, and Andrew Wright. I also thank Jean Cargill, Al Ferry, Rose Treat, and Christal Whelan for useful items of information. I am grateful to the members of the 1994 Mushroom Course at the Harvard Institute for Learning in Retirement for sharing with me their reaction to an early draft.

I thank two exceptional photographers, Charlie Hrbek and Kenneth Kleene, for providing me with beautiful photographs of mushrooms. Many of the figures in the text are the work of John Woolsey, a lover of mushrooms whose fine work has accompanied other projects of mine.

I am especially grateful to Michael Fisher of Harvard University Press, who got so involved with the subject that he became this book's first convert. I appreciate his skill, patience, and good humor at all stages, from agreeing to read my early scribbles to lending a robust helping hand when needed. To thank my editor, Kate Schmit, I must resort to superlatives. Her suggestions were imaginative, thoughtful, sensible, and invariably helpful. I could not ask for greater competence or a more generous engagement with my work. I also thank Kate Brick, who followed in that tradition toward the end of the project.

One of the earliest readers of this manuscript was my daughter, Judith Schaechter. Her keen ear for language, not to mention her warmth and encouragement, helped guide me in my writing. My deepest thanks go to my wife Edith, who lovingly nourished this project with her support, understanding, and finely tuned critical sense.

PROLOGUE ~ THE HUNT

Like most people, I first got interested in wild mushrooms with eating in mind. I was tantalized by the prospect of being able to gather specimens of rare taste scarcely available by other means. I still delight in foraging for edible species, but my horizons have expanded as I have discovered that there is more to mushroom hunting than just looking for those that are good to eat. By now, I have become convinced that a mushroom collector searching only for provisions for the table would be comparable to a bird watcher looking only for quail, ducks, or pheasants. Going on a mushroom walk fulfills all sorts of other yearnings besides the gratification of foraging for natural food. I am excited by the zest of the hunt, challenged by the demands of identification, pleased by the encounter with species that have a special meaning to me, and charmed by especially handsome specimens.

Mushrooms, growing on the ground or sticking out from tree trunks as brackets, seem to color the motif of the forest. Many mushrooms are lovely to look at, varied in hue and shape, smell and texture. As resplendent as flowers, they present us with a range of shades, from forceful brights to subdued pastels. The shiny, lacquer-box red of certain bracket fungi, the violet of the eastern *Cortinarius iodes,* the royal blue of *Lactarius indigo,* or even the spotless white of deadly amanitas feast the eye of the passer-by.

When I lived in Boston, I enjoyed mushroom collecting in the collegiality of a few fellow devotees. Picking wild mushrooms is not the common occupation in North America that it is in most parts of the world where mushrooms can be found in abundance. Still, this is a growth industry: more and more people are getting into mushrooming, especially in northern California and the Pacific Northwest. Because of the abundant rainfall and extensive forests, these are some of the best picking regions on this continent. There seems to be a great deal of latent interest elsewhere as well; I find that "laymen," people who never expressed the urge to look for wild mushrooms before, often respond to my passion with curiosity.

I am often asked how I first became interested in mushrooms, where I go looking for them, and if I have ever been in trouble from eating them. The questions seem to arise mainly from curiosity, but occasionally there is also concern. Some people worry about me, perhaps from an ingrained belief that eating from the wild is a reckless thing to do, akin to keeping poisonous snakes or tarantulas in the house. Many others, however, are truly interested in the topic and ask me, sometimes insistently, to include them in a future mushroom hunt. Often I am told of grandparents, usually of European origin, who used to pick wild mushrooms both here and in the old country and who tried to teach this art to their grandchildren, usually with little success.

Why is mushrooming catching on in North America, albeit slowly? Love of Nature comes in waves. One of the largest of these waves took place at the end of the nineteenth century, when all kinds of outdoor activities, including hiking in the mountains and "botanizing," became not only accepted but even stylish. A more recent wave was part of the social changes of the 1960s and '70s, and we are still feeling its effects. What is more, interest in wild mushrooms parallels the enormous expansion of the American cuisine. In recent years "wild" mushrooms that can be cultivated, such as shiitake or oyster mushrooms, have become commonplace on supermarket shelves. Even truly wild mushrooms—chanterelles, morels, porcini—are dished out (often in pitifully small amounts) in expensive restaurants, where they serve as a mark of culinary refinement. Wild mushrooms are part of the same revolution that brings us readily available cappuccino, sun-dried tomatoes, radicchio, and, miraculously, decent bread.

From enjoying wild mushrooms at the table it is only a small step to searching them out in the forest. (After all, this has appeal for both the foraging instinct and the pocketbook.) More and more people go out in pursuit of edible species, either on their own or with knowledgeable friends. Others do it in an organized way, joining one of the over one hundred North American mushroom clubs, most of which have sprung up in recent years. Once hooked, however, mushroom hunters pursue their avocation with enthusiasm, some with an intense passion. They sometimes talk about mushrooms in a charged, nearly poetic language. Here are the words of Larry Stickney, a San Francisco mushroom lover: "Early in the season, hunting in the cool, magnificent redwood forests . . . can produce both many choice edible mushrooms . . . and an exquisite sense of beauty, tranquillity, and exultation from the deep silence and sheer size of the trees. Right next to a thousand-year-old, three-hundred-foot giant, you can find tiny, fragile, elegant *Lepiotas* and *Mycenas*, which can set your sense of proportion and perspective atingle."

My sentiments exactly. I wrote this book to share these sentiments—for the hunt, for nature, for biology. My expectation is that by telling you stories about mushrooms and the people who study and enjoy them, you too will share my enthusiasm. Mushrooms surprise us in many ways. Do they in fact burst forth "like mushrooms"? Are they friend or foe? All of us know that some mushrooms are food, some are poison, and some alter the mind, but many are not aware of the essential role they play in the perpetuation of life on earth. The study of mushrooms is not an unassuming subject.

Each encounter with a mushroom is a singular event.

Every specimen makes demands of me: Do I know it, and, if not, should I collect it and try to find out what it is? Should I bring it home for the kitchen? Am I to share it with fellow mushroomers? Should I photograph it, sketch it? Should I guess how long it's been around, how long it will last, whether or not it will come back the next week or the next year? What role does this particular species play in human affairs? Has it, or one of its relatives, been used for food, or, in malevolent hands, for poisoning someone? Was it used for altering the mind, for understanding the present, for divining the future?

This book is not a field guide and does not attempt to help you distinguish between edible and poisonous species. To advance in that direction, I suggest that you enlist the help of knowledgeable friends or acquaintances or that you join (or start) a mushroom club. Telling mushrooms apart is not an easy undertaking, especially if all you have at your disposal is a field guide. Should you be blessed with a good visual memory, you may well retain the name-specimen connection, once it is made, for a long time. The more often you go out collecting, the more you are likely to remember. In time, you should develop a warm sense of familiarity with many of the mushrooms you encounter.

In this book I present mushrooms as more than a source of food, even though I find that an interesting and rewarding matter. I am intrigued by how mushrooms have been viewed through time and by different cultures. As a biologist, I find the ways they grow and reproduce unique and fascinating. As a lover of nature, I am fascinated by the different niches in which mushrooms and other forms of fungi are found— ants and termites, for example, have evolved an intriguing interdependence with fungal life. And, lastly, I have found that the people who share my hobby are second in interest only to the mushrooms themselves.

To bring these disparate interests together, I have organized this book to focus on four major topics. The first is the biology and history of the type of organisms that produce mushrooms: the fungi. Next I introduce the reader to the art of mushroom collecting. Then, like many devoted collectors, I have some culinary tales to tell. Finally, I return to the wonder of the biology of the fungi, where the subject is the great variety of nonedible fungi. As poisons, hallucinogens, and the principal crop of ants and termites, fungi have a great many uses both in nature and in human affairs. What a rich subject matter are the mushrooms!

Yoi no ame matsu to chigirite kinoko kana

The evening rain,
And the mushrooms
It promised.

Kisei

Mushrooms and Other Fungi

1

MUSHROOMS IN OUR MIDST

In 1991 the remains of a Stone Age man were found in the Tyrolean Alps. Hikers discovered the frozen and partly mummified body high on a glacier near the Italian-Austrian border. By radioisotope dating, it could be established that the man lived 5,000 to 5,500 years ago. Among his possession were three remnants of tissue from tree-growing bracket or shelf fungi. One of the fragments was almost certainly used as tinder (in fact, the tradition of using bracket fungi for making fire survived in Europe until the advent of matches). The purpose of the other two pieces of fungal tissue is not known, although they have invited speculation that they may have served a medicinal purpose. The finding of three different kinds of mushrooms attests to the importance that fungi must have had in this particular Stone Age society.

Even without this physical evidence, we could safely conclude that prehistoric people would have gathered and eaten mushrooms. Some of the important Cro-Magnon settlements, such as those near the famous Lascaux caves in France, are located in regions that are rich in wild mushrooms and, indeed, renowned for truffles. Mushrooms are not among the subjects of the Cro-Magnon wall paintings, but neither are vegetables of any sort. Then as now, it was necessary to be able to tell edible mushrooms from poisonous ones, but that was a common problem for pre-agricultural society. Every type of root and berry would have to be tested, probably by trial and error, as a source of food. One can imagine the shamans and the older men and women of a clan as an "identification committee" arguing over individual specimens, just as the members of today's mushroom clubs discuss species. How do you suppose they said "What color are the spores?" in Cro-Magnon?

Some ancient societies used mushrooms for shamanistic and divination rites. Evidence for this is suggested by the well-documented practices that have survived to recent times, and even to the present day, in Mexico, Siberia, and New Guinea, for example. In addition, old folk beliefs from nearly all European countries attribute magical and spiritual powers to mushrooms. These remnants of ancient traditions suggest that mushrooms played an important role in the shaping of early cultures, but, appealing and even plausible as this idea may be, the written and pictographic records that support it are limited. Some researchers have found hints of spiritual uses for mushrooms embedded in ancient writings, but, at least for the western world, little explicit documentation exists. Mushrooms are not mentioned unambiguously in the New and Old Testaments or in the Koran. One controversial citation in the Bible is the passage where Elijah (2 Kings 4: 38–40) finds an antidote for poisonous "gourds." The name for these (*paquot*) is the word used in modern Hebrew gilled mushrooms, but it may well have meant "gourds" in ancient times. According to most experts, the biblical description probably pertains to gourds, not mushrooms.

Mushrooms and truffles (*pitriyot* and *kemehim* in Hebrew) are mentioned together several times in the Talmud. In one place it is written that after mushrooms were given to invalids, "some became demented and others died." Elsewhere we find that "people gathered mushrooms and truffles after a rain." According to the Babylonian Talmud (Berakoth 40b), eating mushrooms is not to be preceded by reciting the blessing reserved for vegetables; rather, a more generic prayer is called for: "Blessed art Thou O Lord, our God King of the universe, by whose word everything was created." The reason for this difference is that mushrooms are not ordinary plants because "they do not draw their nourishment from the ground but from the air," which explains why they possess no true roots and "are fed by other plants." This is perhaps the earliest known distinction made between mushrooms and plants. Interestingly, from Aristotle to well into the twentieth century, the western world steadfastly believed mushrooms to be plants. (As I will explain in the following chapter, a mushroom is one of many forms a fungus may take. That is why the words *mushroom* and *fungus* are often used interchangeably.) Various "primitive" peoples in several parts of the world have had no trouble separating mushrooms from plants, however. For example, in the Mexican state of Michoacán, Indians hold the view that "mushrooms are not plants, they are flowers of the earth."

The earliest extensive references to mushrooms to be found in the western literature come from classical Greece. Fungi appeared in the writings of Euripides and Hippocrates in the fifth century B.C., both of which mention several cases of mushroom poisonings. There is ample reference to the joys and dangers of eating mushrooms in the writings of Greek and Roman philosophers, naturalists, and physicians, such as Theophrastus, Pliny the Elder, Galen, and Dioscorides. Very little written material is available, however, on the psychedelic effects of mushrooms or on their spiritual powers. The Greeks and Romans were appropriately perplexed about the origin of mushrooms and seem to have given much thought to the matter. They were puzzled by the fact that mushrooms do not have visible "seeds" and seem to appear out of nowhere. The best explanation they could come up with is that mushrooms are excrescences of trees, ferments of the earth, or the products of thunderbolts. According to Suetonius, the Greeks called mushrooms the "food of the gods," but the Greek philosopher Porphyry preferred to call them "sons of the gods," because they are "born without seeds."

The Romans held certain mushrooms in the highest esteem and served them at their most sumptuous banquets. Special dishes and implements were used for their preparation. The Romans considered one kind so exquisite that it merited being called "Caesar's mushroom"; it retains the appellation *Amanita caesarea* to this day. This may be the mushroom that Martial invoked in an epigram: "Silver and gold and a fine cloak—these are easy to send with a messenger. To trust him with mushrooms—that is difficult!" It also appears that the Romans used mushrooms directly or indirectly for political murder—for instance, in the poisoning of the emperor Claudius.

Mushrooms are mentioned as delicacies in European cookbooks of the Middle Ages, but relatively little scholarly references come to us from this period. Avicenna (979–1037), the great Moslem physician, gave some advice regarding cures for mushroom poisonings and warned that black and green mushrooms and those the color of peacocks were to be avoided. He may have meant *Amanita phalloides,* the great killer among mushrooms, which often has a greenish tinge.

By and large, the scholarly literature on mushroom biology and edibility that has survived from early modern Europe is confusing and full of contradictions. One surmises that ordinary people knew more on this subject than the scholars. Albertus Magnus mentions in his thirteenth-century book *Of the Plants* that the fly agaric (*Amanita muscaria*) was used to kill flies by northern tribes, a practice that has been well documented since then among people of many parts of Europe and Asia. The modern era of mycological study did not begin until the Renaissance. The fifteenth-century Venetian naturalist Barbaro Ermolao (latinized as Hermolaus) was the first to revive, revise, and expand the botanical knowledge of classical Greeks and Romans. In the daring spirit of the times, Hermolaus critically reviewed the entire works of Pliny the Elder in his *Castigationes Plinianae*. Hermolaus and other scholars of the fifteenth and sixteenth centuries left a small but stimulating collection of mycological writings, which has served as the wellspring of knowledge for later studies.

The earliest published treatises on mushrooms were modest undertakings. Publishing lengthy books was economically demanding in the Renaissance, but short treatises, or "dissertations," were popular. It is not surprising, therefore, that the first publications in this field were pamphlets that dealt specifically with one kind of mushrooms. The very first one appeared in 1564, written by the Dutch physician Hadrian de Jonghe (1512–1575). It was thirteen pages long, included a wood cut, and dealt with the phallic-looking stinkhorns (Chapter 9). The medicinal virtues of these mushrooms for treating gout and arthritis were extolled in prose and poem. One species of these mushrooms is appropriately called *Phallus hadriani* to this day. The same year, *Opusculum de tuberis* ("A Pamphlet on Truffles") was published by the Italian Alfonso Ciccarelli. We don't know this author's exact birth date, but the date of his death can be traced with precision because he was hanged in 1580. (His execution was not provoked by his writings on mushrooms, though; Ciccarelli—perhaps too much of a Renaissance man in the breadth of his interests—was caught falsifying genealogies and titles of nobility.)

Speaking of nobility, mushrooms even appeared in coats
of arms. I came by this fact by bringing up the subject with
a Danish friend, Jens Ole Rostock, who was an expert in
heraldry. He promised to investigate the matter and later
wrote to me that he "went through the keys of about 150,000
European coats-of-arms and found only 18 with mushrooms.
Of these, 10 are French, 3 German, and one each English, Belgian, Swiss, Italian and Polish. This work, of going
through the keys of that imposing number of coats-of-arms,
is by no means as gargantuan as it may sound, since heraldic
keys are arranged in exactly the same manner as botanical
ones." Jens Ole found that mushroom names were sometimes used as puns for the family name, which is an established heraldic tradition. Thus, mushrooms appear in the
arms of the Fongarini, the Lesseps (a pun on "les ceps"), the
Boulet (which sounds like "bolet"), and the Moreau (which
reminds us of morels). The Rabasse arms is straightforward,
since *rabasse* is Provençal for truffles, and so is the symbol
of the Pilz family (*Pilz* is the German word for mushroom).
See Figure 1.1.

The model of the Renaissance mycologist was Charles de
l'Ecluse, known by the latinized name of Clusius. He was
born in the Netherlands in 1526 and lived for eighty-one
years. A lawyer, a philosopher, and a physician, Clusius
became deeply involved in the theological issues of his time.

1.1
A sixteenth-century
German coat of arms:
the Kreis family.

His ardent interest was botany, to which he contributed the discovery of many new species and writings on the acclimatization of cultivated and decorative plants and the foundations of ecology. He traveled extensively in Europe and wrote a number of books on the plant life he encountered. Among his works is *Fungorum in Pannonis Observatorum Brevis Historia* (1601), or "A Brief Inquiry into Fungi Found in Pannonia," the first extensive monograph on mushrooms ever published. Pannonia, I found out, comprises the western half of Hungary and parts of the adjacent countries.

The "Brief Inquiry" contains detailed descriptions of over twenty-five mushroom groups. Most of the species can be identified with assurance from the illustrations and from their descriptions, which include both Hungarian and German common names. Many of these names are used in that region to this day. To illustrate his account of Pannonian mushrooms, Clusius commissioned watercolors of many of the specimens. It is not known with certainty who the artist was, but it may have been his nephew, Esaye le Gillon. These watercolors were believed to be lost even in Clusius's time, but they reappeared in the nineteenth century in the library of the University of Leyden. Clusius had no means to publish the watercolors, since techniques for color printing were not available at the time. The eighty-seven watercolors, which are known collectively as the "Codex Clusii," are remarkably well preserved and have lost little in brilliance over the 400 years of their existence.

At the time Clusius was studying the natural history of

mushrooms, others provided mushroom eaters with concrete advice about edibility. John Gerard, the English author of a famed herbal (1597), printed a number of drawings taken from Clusius and made what now appear to be arbitrary statements about their edibility. To explain why some mushrooms are poisonous, he quoted Dioscorides: "poisonous mushrooms groweth where old iron lieth or rotten clouts or neere to serpents dens, or roots of trees that bring forth venimous fruit." Gerard relied on the most concise classification of mushrooms possible: "The Mushrooms or Toadstoole . . . hath two sundrie kinds, . . . for the one may be eaten; the other is not to be eaten."

Already in the sixteenth century, the English manifested the fear of mushrooms that until recently has distinguished them from other Europeans. Gerard apparently shared in this opinion:

> Therefore I give my advice unto those who love such strange and new-fangled meates, to beware of licking honey among the thornes, lest the sweetnesse of the one do not countervaile the sharpness and pricking of the other . . . To conclude, few of them are good to be eaten, and most of them do suffocate and strangle the eater.

Mycology entered the truly modern era in the following century, with studies that eventually elucidated the mystery of the life cycle of many kinds of fungi. Perhaps the most crucial contribution was that of the Florentine Pietro Antonio Micheli (1679–1737), whose work helped settle a long-standing controversy: do mushrooms reproduce by seeds (or some such), or do they arise spontaneously? Micheli made the first reliable and detailed microscopic observation of spores of mushrooms and molds and conducted a series of stellar experiments proving that fungi develop from spores. He picked spores from mushrooms by using a very fine brush and sowed them on leaves; with mold spores he chose sections of squash as the substrate. He found that the fungi reproduced their own kind from these "seeds." Following a procedure that was later to become a basic tenet of microbiological technique, Micheli went one step further and "re-inoculated" a new section of squash with spores from his first culture, getting the very same mold to grow once again. On one occasion, he inoculated one part of the squash with a black mold (the bread mold *Mucor*), another part with a green mold (a common *Aspergillus*), and showed that the different spores were able to "breed true." This was one of the first experiments to cast doubt on the theory of spontaneous generation, which was finally laid to rest by the work of Pasteur and others in the nineteenth century.

Micheli's research, although much noticed, was not universally accepted at the time, in part because attempts to repeat his results with other fungi often failed. This is not surprising, however; even today some species of fungi are difficult to cultivate from spores.

Micheli, like many later mycological researchers, was not a university professor or an academic in the usual sense. Born to modest means, Micheli as a young man became the botanist to the court of Duke Cosimo III of the Medici family. This post allowed him to carry out his work for some years, albeit under austere conditions. His fortunes changed for the worse in his latter years, and he died in penury. He left a most impressive body of mycological works, including important contributions to the taxonomy of the fungi. An interesting diversion in Micheli's life was his experience as an "industrial spy." His patron wanted him to appropriate the secret of the manufacture of tin plate, which had been recently discovered in Bohemia and which had become a German monopoly. Why this task should be entrusted to a botanist is not clear, but it is known that Micheli tried to disguise his true intent by gathering local seeds and plants. His foray into metallurgical espionage came to naught, but at least he benefited from his botanical "cover."

The cataloguing and classification of mushrooms continued in the eighteenth and nineteenth centuries, but scientists worked in an independent and individual manner that led to considerable confusion. The person credited with creating order from chaos was Elias Fries (1794–1878), a Swede, who compiled much of the available knowledge into a coherent and readily usable scheme. His classification of mushrooms, based largely on characters visible to the naked eye, appeals both to the senses and to common sense. It has come down to us as the "Friesian classification," but it has slowly been superseded within the last 100 years, when the widespread use of the microscope introduced the use of new taxonomic characters. At present, mycological taxonomy is about to enter a new era, thanks in good part to the availability of molecular techniques for determining the kinship of different mushrooms. The early returns from work with DNA analysis tell us that much of the current taxonomy, based largely on morphology, holds up pretty well. Let us hope that things don't change too much. It's hard enough for the amateur mycologist to remember one set of names without having to learn a whole new nomenclature.

The goal of the science of mycology, of course, is an understanding of all aspects of every kind of fungus. Those of us who are not professional mycologists tend to take a more limited approach to the field. Nevertheless, people almost everywhere have definite feelings about mushrooms, and few are those who are indifferent to them. I used to think that people who live in temperate zones would be the most avid mushroom hunters, but it turns out that this is not so. Many people who live in the tropical regions, especially in Africa and Asia, regularly gather and consume many varieties of wild mushrooms. Even in the desert, all the way from Morocco to Iraq, people harvest several kinds of "desert truffles" and prize them as valuable food.

In most places of the world, the attitude toward mushrooms is generally positive, ranging from mild interest to unbridled enthusiasm. There are, however, a few parts of the world, most notably the British Isles, where the more common attitude is disdain and even fear of wild mushrooms ("mycophobia"). Most other Europeans, from northern Spain to Russia and from Italy to Scandinavia, are mushroom lovers, "mycophiles." In these parts of Europe, people are not in awe of wild mushrooms because harvesting them is a common experience, especially during childhood. Looking for mushrooms is taken for granted, something that is part of a walk in the woods in the autumn. Even in Europe, however, someone like me belongs to a minority because I am interested in *all* mushrooms, not just the edible ones. Most Europeans don't seem to understand what could be so fascinating about mushrooms that one cannot eat. What an odd thing to do! In German there is even a term—*Pilznarr,* or "mushroom fool"—that describes me and other addicts of the passionate study of all mushrooms.

One would think that in places where mushroom lore and culinary uses are widespread, people would be particularly careful about what they eat and what they avoid. Not true: the more people collect wild mushrooms, the larger the number of cases of poisoning. Of course, this may simply be due to the larger number of opportunities for poisoning. On the other hand, there is ample evidence that familiarity with mushrooms breeds conceit, sometimes with disastrous consequences. The gifted composer and famous organist and harpsichordist at the court of Versailles, Johann Schobert, is an example. In 1767, Schobert picked wild mushrooms with his family in Pré-Saint-Gervais, a village near Paris. They took the harvest to a nearby restaurant and asked to have the mushrooms prepared and served. The chef examined them and told Schobert they were poisonous, and the party

Mushrooms Shed Light

Occasionally a mushroom hunter will encounter a bright orange beauty, the "Jack O'Lantern mushroom," growing in conspicuous clusters, usually around the base of hardwood trees or stumps. It can be so shiny and resplendent that it seems an especially welcome find. Alas, it is a poisonous kind, given to induce gastrointestinal afflictions and profuse sweating. Poisonings from these mushrooms are not unheard of, because beginners have been known to confuse them with chanterelles, with which they have a superficial resemblance. There are two species of Jack O'Lanterns in North America, *Omphalotus olivascens* on the West Coast and *O. olearius* on the East. As if trying to compensate for being inedible, the Jack O'Lanterns are an unexpected source of *visual* delight: they emit light. Along with some other mushroom species, they join fireflies, glowworms, and certain marine bacteria in being *bioluminescent*.

The greenish light, known as "foxfire," is given out not only by the mushrooms themselves but also by the mycelium, the fungal filaments that often permeate the wood of diseased trees. The surface layers of the mycelium of such impregnated wood, which is called "touchwood," can be seen to glow fairly brightly for one or two weeks. This property has inspired fear and wonder since time immemorial. Imagine finding a tree branch shining bright (with apologies to William Blake) in the forests of the night! Legends describing such eerie encounters can be

continued

found in ancient Greek, Roman, and Indian texts. It has even been suggested that this phenomenon may explain the biblical story of the bush that burned without being consumed, showing Moses the way to the Promised Land. It was pointed out by the British mycologist John Ramsbottom, however, that Moses was unlikely to have led the way at night, when the luminescence would be visible.

The whole subject of bioluminescence is wanting for an explanation. Luminescent animals may conceivably use the light to find mates or food, but this can hardly be the reason mushrooms glow in the dark.

People from many parts of the world have found uses for these natural lanterns. The Swedish historian Olaus Magnus wrote in 1652 that people in the far north of Scandinavia would place pieces of rotten oak bark at intervals when venturing into the forest. They could then find their way back by following the light. Similarly, during World War I soldiers in the trenches placed touchwood on their helmets to keep from bumping into others in the dark. The Native American herbalist Keewaydi-noquay relates that an Ahnishinaubeg shaman of her acquaintance positioned two glowing wooden pillars on either side of her doorway, much as suburban homeowners arrange lights on a front lawn. These ghostly lights scared visitors instead of attracting them, however, and the logs were soon dumped.

left in a huff. They tried again, however, this time in a restaurant in the Bois de Boulogne, with the same result. Apparently Schobert got quite angry because a doctor in the party had declared that the mushrooms were edible! Finally, at Schobert's home, they prepared a soup with the mushrooms. Schobert, his wife, all but one of their children, and the doctor all died. Schobert was then in his late twenties or early thirties (the date of his birth is not known with certainty), and his death deprived us of a talent said to rival that of Mozart.

That the British, living in a mushroom-rich region, should be so particularly mycophobic has puzzled many writers, but I have not yet read a truly satisfactory explanation. In the middle of the last century, the Reverend Charles Badham, the author of a fine mushroom field guide, noted that "England is the only country in Europe where this important and savoury food is, from ignorance or prejudice, left to perish ungathered." Anglo-Saxon mycophobia should not be overstated, because a few kinds of wild mushrooms—horse mushrooms (*Agaricus arvensis*) and blewits (*Lepista nuda*)— have been sold and are still being sold in British markets, especially in small towns. Of course, mycophobia exists elsewhere, too. Mushrooms are even shunned in areas where they could be put to especially good use. Larry Millman, a travel writer and mycologist, told me that during his visits to the Canadian Arctic he was amazed that the local Inuit scorned the abundant mushrooms that appeared in the tundra during the summer. This seemed particularly odd in view of the scarcity of food in the region. The simplest way to explain mycophobia of this sort may be that it represents an atavistic taboo derived from some early mycological misadventures of the clan.

Mycophobia has taken root all over North America. It is not uncommon for people here to regard wild mushrooms with considerable fear and mistrust. Many more people now eat the exotic mushrooms available at food stores, but most still look at wild mushrooms with suspicion. Although many Native Americans had a variety of uses for mushrooms, both as edibles and for their medicinal power, the Pilgrims did not seem to develop the taste for mushrooms that they came to have for other native food. Neither did the Spaniards and Portuguese, and, consequently, mushrooming is not very popular in most parts of Latin America either. Traditional mushroom picking is confined to areas where Native Americans are still able to carry out their customs, such as southern Mexico, Guatemala, and the Amazon basin. A few other pockets of earnest mushroom consumption exist in regions were there is an abundance of edibles, such as southern Chile and Argentina.

Nowhere else does wild mushroom picking reach the level of passion that is seen in Russia. There, *hodit' po gribï* ("looking for mushrooms") approaches a national craze, and few are those who do not participate in the hunt. Older people remember that during World War II, when gruel or potatoes were often the only food available, wild mushrooms were a greatly appreciated addition of flavor. Even now, expeditions (sometimes requiring special trains and buses) are frequently organized in the fall to deliver Moscow's aficionados to choice sites. There is more to the mushroom craze than meets the eye, as Russian urbanites often use mushroom picking as an excuse to spend time in the countryside, cook over an open fire, and choose a different type of vodka to suit the species of mushroom just collected. Even in Russia, however, most people know only about a dozen edible species and shun the rest. In 1967, the popular Russian novelist Vladimir Solouhin wrote a widely read book that extolled many more kinds of mushrooms. It seems that his warnings about the danger of eating unidentified mushrooms went unheeded; so many people became ill that they were called "Solouhin patients."

Various mushrooms—such as shiitake, paddy straw mushrooms, oyster mushrooms, and many others—are consumed in great quantities in Asian countries. In Southeast Asia, mushrooms are part of the daily diet, and the demand there exceeds what can be gathered naturally. This has led to widespread cultivation of many mushroom species, far outnumbering the few that are commonly raised in the West. In Japan, hardly a day goes by that a person does not eat a dish containing at least some mushrooms. One or two caps of shiitake invariably appear in a noodle dish, glutinous nameko in a soup, enoki as a garnish in a salad. Picking wild mushrooms, however, is not that widespread an activity in Japan; most people confine themselves to buying mushrooms in the grocery store. I have wondered if this relative indifference to wild mushrooming may not be due in part to the intimacy of the Japanese with the cultivated varieties. After all, who would go hunting for seaweed? One kind of wild mushroom, however, stands apart for the Japanese—matsutake, or pine mushroom. Matsutake is appreciated more for its aroma than for its flavor. In the past, this was *the* prize that many people eagerly sought in the pine forests. Today, matsutake makes a fine gift for one's boss on a special occasion.

For all we know, mushroom eating may be an ancient biological attribute of humans since it is also a trait of our evolutionary cousins, the mountain gorillas. Dian Fossey, who studied gorillas in the wild for decades, found that a species of bracket fungi was eagerly consumed by the gorillas she studied. The fungal species is called *Ganoderma applanatum,* also known as the artist's fungus because the whitish undersurface of the shelf becomes dark on touch and can be used for drawing (although I suspect that the gorillas are indifferent to this fact). Fossey writes in her book, *Gorillas in the Mist:*

The shelflike projection is difficult to break from a tree, so younger animals often have to wrap their arms and legs awkwardly around a trunk and content themselves by only gnawing at the delicacy. Older animals who succeed in breaking the fungus loose have been observed carrying it possessively from more dominant individuals' attempts to take it away. Both the scarcity of the fungus and the gorillas' liking of it cause many intragroup squabbles, a number of which are settled by the silverback, who simply takes the item of contention for himself.

Except for different table manners, this description does justice to the affinity that some of us have for wild mushrooms.

2

THE LIFE HISTORY
OF THE MUSHROOM

"Hold That Fork! The mushroom in your salad could be a (very) distant relative." This headline, in *Time* magazine (April 16, 1993), introduced a story reporting that fungi are more closely related to humans than to plants. Mitchell Sogin of the Marine Biological Laboratory at Woods Hole, Massachusetts, had recently published an article in the journal *Science* on the genetic relationship between protozoa, sponges, fungi, animals, and plants. A parallel study by Jeffrey Palmer and Sandra Baldauf had also appeared that year. These researchers deduced that a single-celled ancestor of both fungi and animals may have split off from the lineage that led to plants about 1.1 billion years ago.

This seems counter-intuitive to most people—after all, the type of fungus most familiar to us, the mushroom, seems so much more like a plant than an animal, doesn't it? Perhaps the Yanomamos of the Amazon would not have been surprised, however. These inhabitants of the rain forest have one word for eating meat *and mushrooms* and another for eating everything else. This may simply attest to the fact that mushrooms are a good substitute for meat, but might it also reflect a deeper understanding of their kinship?

I will leave questions about ancestral lineages to evolutionary biologists. More pertinent to someone with an interest in mushrooms (and mushroom hunting) is their taxonomy—their place in the classification of all living things. Taxonomists assign all organisms to established categories, from the largest division (the kingdom) to the smallest (the species). Although there is still some argument over the details, most biologists will agree to five major kingdoms of life: Procaryotes (single-celled organisms lacking a true nucleus, such as bacteria), Protista (single-celled and other microscopic organisms with a nucleus), Fungi, Plantae, and Animalia.

As you can see, mushrooms belong to a kingdom of their own. Also referred to as the Mycota, this kingdom includes not only the mushrooms but also molds, yeasts, smuts, and mildews. The study of these organisms is called *mycology,* from the Greek *mykes* (for "fungus"), but the different kinds of fungi share more than a name. All fungi differ from other organisms in a number of biological properties, the main one being that fungi need organic matter made by other living things for their growth and sustenance. Unlike plants, they cannot use photosynthesis to produce their own food from carbon dioxide, inorganic materials, and light. It is in this way that fungi most resemble animals: both ultimately depend on plants for food.

We do not really know at what point in evolutionary history the fungi split off from the animals and evolved as a separate group. The most reliable information concerning the origins of fungi comes from fossil fungi found inside plants. Trees appeared on earth some 300 million years ago, and fungi have been discovered in fossilized wood. This goes to show that the association of fungi and plants is an ancient one.

What is seen as fungal remains inside plant tissue is in the form of filaments, but these filaments reveal little about the identity of the fungus. Whole mushrooms have very rarely been found as fossils, as anyone who has ever kept mushrooms in the refrigerator too long might expect. However, a few fossil specimens *are* known: there is a Patagonian bracket fungus from the Jurassic period, and a tiny cap-and-stem mushroom from the Dominican Republic has been found to be about 20–35 million years old. A particularly exciting find was the discovery of two amber-encased tiny specimens from central New Jersey. One of them appears to be whole: a cap about one-eighth inch across with a dozen or so visible gills and a stem, looking very much like a typical mushroom of today. These specimens, from the Cretaceous, are about 90 million years old, suggesting that some of the mushrooms alive today may have a very ancient origin.

In other words, the makers of the movie *Jurassic Park* could have put some mushrooms on their island hideaway. The sci-

Mushrooms in Jurassic Park?

continued

entists could have extracted DNA from mushrooms encased in amber, as they "mined" DNA from fossilized insects. Then, according to the scenario created by Michael Crichton, they would have used techniques such as PCR (polymerase chain reaction) to amplify the minute amounts of DNA. Even if this procedure does not allow the cloning of the whole organism (despite the plot of the book and movie), it would produce enough genetic material for analysis to establish taxonomic relationships and tell us something about how far the fungi have wandered over the course of time.

This is not to say that fungi are chemically inactive. To the contrary, they are credited with a variety of well-known activities, such as making bread rise, causing the fermentation of beer and wine, ripening and flavoring cheeses, and producing some of the most useful antibiotics. Fungi cause infections of the skin, the mucous membranes, and deep tissues of humans and animals. Fungi are also among the most important agents of plant disease: one causes the Chestnut Blight, another the Dutch Elm disease. A kind of fungus was responsible for the great Irish potato famine of the nineteenth century. In a variety of ways, then, fungi have had a huge impact on human affairs.

In fact, though we often don't notice them unless we or

the crops are infected, fungi make up a significant proportion of all living matter. I could not find a calculation for the weight of all the fungi on earth, so I tried my hand at an estimate. I used a conservative figure for the amount of fungi per square meter of soil that supports vegetation, 100 grams (less than a quarter of a pound), and multiplied it by the total area of such soils. I came up with the figure of 10 trillion kilograms of fungi on earth. This figure is certainly an underestimate because it does not include the lichens (part fungi and part algae) that colonize many trees and rocks, save those in extreme deserts and permanently sub-zero polar regions. Even this low count leads to the conclusion that at the very least there are two tons of fungi for every one of us.

Perhaps the greatest role fungi play, in the global sense, is that of major recyclers. Fungi are responsible for most of the recycling of vegetable matter, from a blade of grass to a large tree. They are the great decomposers, able to break down not only cellulose but also lignin, the main component of wood. Even termites and other wood-eating insects depend on fungi and bacteria for decomposing otherwise indigestible compounds. Were it not for the fungi, dead plants and trees would accumulate to great depths and become a "sink" in the carbon cycle. In time, there would not be enough carbon dioxide to sustain plant photosynthesis at more than a snail's pace, and animals that ultimately depend on plant life could not survive. So important are fungi in the recycling of organic matter that life on earth as we know it would be impossible without them. Even so, the fungi are often ignored. For one thing, the recycling of natural matter is a relatively new concern. To some extent, biological degradation has been of secondary interest even among scientists, perhaps because human beings are more attracted to growth than to decay. We think of scavengers in negative terms and do not make allowance for their essential service in the cycles of matter.

Another reason the fungi often go unrecognized is that they tend to be much less conspicuous than plants or animals. Most fungi are microscopic; they exist as filaments in soil or in decaying tree trunks and other vegetation. They are visible to the naked eye only under special circumstances, such as when mold accumulates on spoiled fruit or a smut or rust grows on a plant. At one stage of the reproductive cycle, however, some fungi produce fruiting bodies that are very noticeable. At this stage, the fungus is what we call a *mushroom*.

Mushrooms seem to spring from the ground overnight, appearing where there was none before, challenging our senses and fostering suspicion. The fact that we say "to mushroom" to mean "to develop explosively" illustrates our recognition of this impetuousness. Not only do mushrooms appear suddenly, but for a long while their origin was a mystery as well: they have no seeds, no spreading roots, none of the devices used for reproduction by plants. Until it became known in the early eighteenth century that they reproduce via microscopic filaments and spores, any guess about how mushrooms arise was as good as any other. In times past, mushrooms were believed to result from the decomposition of organic matter, or to emerge after bolts of lightning hit the ground. Surprisingly, the "thunderbolt theory" was held by peoples as diverse as the ancient Greeks and Romans, the Mayans, and the Filipinos.

Mushrooms do indeed develop rapidly, but close study has revealed their secrets. The mushroom one sees aboveground or stuck to a tree trunk is just the visible part of an organism that grows unseen until its time for sexual development has come. The organism consists of microscopic filaments, or cells, called *hyphae* (singular: *hypha*), which may spread out over a considerable expanse, covering square yards or even acres. Because the filamentous growth is not readily detected, there is a gulf between what our senses tell us and what really goes on.

Fungi, along with algae, mosses, ferns, and other "lower" forms of life, are *cryptogams,* which roughly translates as "having hidden sex." The flowering plants, in contrast, display their sex openly. In older days, when the taxonomy of living things was of more general concern, universities had chairs of "cryptogamic botany." Witticism abounded regarding the origin of the term: did the "cryptic" part of the name refer to the seemingly indecipherable aspects of biology, or to the obscure ways the professors had of expressing themselves? I regret the term is now in disuse, since it had a nice enigmatic sound to it.

In "hidden sex," there are no sperm and ova, or seeds; there are spores. Picture a spore newly released from the gill of a mushroom and carried aloft by breezes. If it lands in a suitable site in the woods or a field, it will sprout and grow to form a single microscopic filament, a hypha. If the food and water supply is adequate, the filament will make multiple branches and elongate in every direction. In the complex environment of soil, each living thing competes with others, and these fungal filaments are no exception. If they are successful, the filaments will grow into a loose, cottony mass, somewhat akin to the "fuzz" that covers decaying fruits or breads. This accumulation of hyphae is called the *mycelium,* but it is not readily visible as a structure in nature because it is usually interspersed among particles of soil, fallen leaves, or the woody tissue of a tree.

A mycelium may grow to occupy a spectacularly large area. Reports from Michigan tell of a single mycelium of a variety of the honey mushroom *(Armillaria bulbosa)* that covered some forty acres of land. This enormous biomass represented a single individual; that is, each hypha was genetically identical from one end of the mass to the other. Researchers at the University of Toronto and Michigan Technical University claimed that such a mycelium is not only among the largest of all living things but is also one of the oldest, about 1,500 years of age. However, as was pointed out by Gary Lincoff, a New York mycologist, similar claims can also be made for grasses and those trees, such as aspen or elm, that grow by extension of a single root system, without the benefit of seeds. Still, a mycelium can occupy a very large area, and sibling mushrooms may arise from the same mycelium over apparently separate parts of a forest or pasture.

Much more than just growth of the mycelium needs to take place before a mushroom can be formed. The filament that arises from a single spore will not usually become a mushroom spontaneously; it will just grow larger and branch into an ever-spreading mycelium. The formation of a mushroom requires the onset of the sexual cycle. The cycle begins when one fungal filament pairs ("mates") with another filament (pairing here means the physical fusion of two filaments). The two filaments must be of different "mating types," or genders, and at this point the story gets complicated. The actual number of genders or mating types in certain species is very large, which makes the number of possible pairwise combinations astounding. (See Figure 2.1.) In one well-studied species (the "split gill," or *Schizophyllum commune*), some 21,000 kinds of pairings are possible! What makes it difficult to study this process is that filaments of all genders look alike and can only be distinguished by their mating behavior. On the other hand, *they* know the difference!

Mushrooms do not have sex organs—in principle, any filament is poised to participate in mating and does not require specialized structures. Many (but not all) mushrooms have not just two genders but many more than that. The notion of such sexual variety takes some getting used to, but the basic rule is the same for fungi as it is for humans: to produce a viable offspring, mating must take place between individuals belonging to different genders. Mating with the same gender is infertile. We don't know what to call all the various genders, as we only have names for male and female. (How about a parlor game whose purpose is to invent the names for a couple of dozen other genders? Just in case we end up evolving like the fungi?)

Explaining how such a large number of fungal genders comes about is where the subject gets technical. In animals and higher plants, each gender is defined by a *single set of genes* that comes in two versions, one for maleness and one for femaleness. Each set does not consist of a single gene but of a complex of many genes acting in unison. In many (but not all) of the higher fungi, each mating type is determined by *two* sets of genes. Each set probably specifies ("encodes") distinct steps in the mating process. For example, one may be responsible for the fusion of the filaments, another for the behavior of their nuclei. As of now, we know little about how these genes operate, but we do know that two sets of genes are present in every individual spore and, therefore, in the filament it produces. Unlike in the human genome, these genes can have hundreds of individual variations and *each pair of variants specifies a different mating type or gender.* In *S. commune* mentioned above, for example, if we call the one set of genes A and the other B, we know of 350 variants of set A and 60 variants of set B, which can combine in $350 \times 60 = 21,000$ different ways.

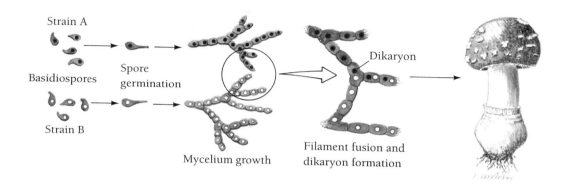

Strain A

Basidiospores

Spore germination

Strain B

Mycelium growth

Dikaryon

Filament fusion and dikaryon formation

2.1
How to make a mushroom: Fungal spores of different mating types (shown as having white or black nuclei) germinate and form filaments (hyphae). These fuse together to make a new filament with *both* kinds of nuclei. The dikaryon may eventually differentiate to produce a fruiting body, what we call a mushroom.

Don't be misled into thinking that A stands for maleness and B for femaleness, rather than for two different sets of genes. The distinction, once again, is that each individual has both A genes and B genes, not just one of the sets. The point is a subtle one, and indeed it is often misinterpreted. John Cage, the late avant-garde composer and also an avid mushroomer, misstated the case in an interview: "There are around eighty types of female mushrooms and around one hundred and eighty types of males in one species alone . . . This led me to the idea that our notion of male and female is an oversimplification of an actual complex human state." Cage may well have been right in his conclusion even though he oversimplified the details. To me, the existence of more sexual genders than just males and females seems at least as intriguing.

Why is sex among the fungi so intricate? One way living things gain an advantage in the struggle for survival is by "outbreeding," mating with related but different forms. In this manner, organisms acquire new genetic potential and thereby increase their chances to withstand external challenges. They can thus become better adapted to new environments and colonize new habitats. With so many genders, the fungi maximize their ability to breed with new individuals. With the possibilities being so large in number, gender becomes practically irrelevant. This means that in the real world, in the soil or a dead tree, a filament of a given species can mate with nearly any other filament it encounters, because the chances of running into a filament like itself are limited. Contrast this with our own situation: a man or woman can establish a fertile union with only half the members of the species.

Beyond the initial pairing, fungal development has yet other biological peculiarities. The actual fusion of fungal filaments resembles the combination of spermatozoa and eggs, no matter how complex the events leading up to it. But, once again, there is a difference. In the higher forms of life, the nuclei of the two parental cells fuse in the fertilized egg into a single nucleus. In each parental cell there is a single set of chromosomes (a *haploid* set), and these combine in the new nucleus to make a double *(diploid)* set. In humans, this is the first step in the development of the embryo. In the fungi, the fusing of the two nuclei into one happens long after mating. When two sexually compatible filaments mate, the nuclei remain separated (haploid). They remain so as the filaments grow, up to the moment when a fully mature mushroom forms its spores. Thus, for most of its life, each filament contains two separate kinds of nuclei. Cells with two nuclei are known as *dikaryons.*

Why do the cells wait until the last possible minute to combine their nuclei? At first glance, combining nuclei early or late would seem to make little difference: in each case, two sets of chromosomes are carried by the same cell. If the nuclei fuse, however, they form an irreversible union that seals the fate of their (now diploid) genetic material. The dikaryon, on the other hand, by keeping the two nuclei apart, provides another shot at genetic diversity. In the words of a Boston mycologist, Richard Batchelor, the two nuclei live as roommates, capable of further dalliances. A dikaryotic filament can still fuse with another filament, leading to a new combination of nuclei and to a different arrangement of genes. It seems that fungi wish to leave themselves open to new genetic opportunities until the last minute. If we applied the terms of human behavior to mushrooms, we would conclude that for the fungi "commitment" is highly tentative.

The stages of the sexual cycle described so far all go on underground, or within a decaying log, removed from human eyes (unless the natural process is mimicked in the laboratory and observed under a microscope). What triggers the initiation of that part of the cycle we see—formation of the mushroom? As any mushroom picker knows, moisture is essential. Mushrooms often arise quickly, sometimes two or three days after a good soaking rain, suggesting that the filaments were present already in the substrate and that it was moisture that triggered their differentiation into a mushroom. Many mushrooms seem to respond to the same stimuli, which explains why different species arise more or less simultaneously in the same corner of a forest.

Besides moisture, what stimulates the development of a mushroom? In the laboratory, one of the better-known stimuli is, oddly, deprivation of food. When there is no food available for the filaments to grow, do they dedicate themselves to reproduction? The mycelium is capable of undertaking the great changes required to begin the cycle anew because it's loaded with previously stored nutrients. The effect of food deprivation may be mediated by the synthesis of chemicals that trigger the development of the mushroom. Not all cultivated mushrooms require starvation of the mycelium, but many depend on it. The exciting recent breakthrough in growing morels was based on the discovery that the growing mycelium needs to be starved in order to undergo the changes that lead to the formation of fruiting bodies (Chapter 8).

What can one observe when a competent mycelium is on its way to becoming a mushroom? The first noticeable event is that filaments that used to make up a loose aggregate now form a compact latticework. This assemblage, known as the *primordium,* becomes a small knob that can barely be seen with the naked eye. (See Figure 2.2.) In time, this aggregate grows and eventually reveals the shape of a tiny but well-defined mushroom. (Many types of fungi, such as yeasts, never form mycelia or mushrooms, but their various means of reproduction are beyond the scope of this book.) Because the shapes of both the cap and stem emerge so gradually, a renowned British mycologist, E. J. H. Corner, compared their appearance to that of the Cheshire cat in *Alice in Wonderland.*

The baby mushroom now grows rapidly. Most of the increase in size is due not to the division of its cells but to the uptake of water and extension of the individual cells in one dimension. This explains in part why mushroom development seems so abrupt, since the absorption of water occurs very rapidly. The various parts of the mushroom—cap, stem, gills—grow in an orderly fashion, but little is known about how their development is coordinated. It seems that the gills, which are formed rather early, are the site of synthesis of hormone-like regulatory substances that tell the cap and the stem when to expand. This would make sense because the gills, being the site of spore formation, are, in a way, the "business end" of the mushroom—they might as well function as the strategic center that dictates when and how the rest of the structure is to be formed. The new spores, dispersed by the wind, then form new filaments, bringing the cycle of life full circle.

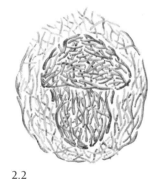

2.2
The earliest indication of a mushroom-in-the-making is seen when a tangle of filaments becomes organized and differentiated.

A fully formed mushroom is a marvel of structural intricacy and a pleasure to behold. To learn something about the anatomy of a mushroom, you can start with the cultivated kinds in the kitchen or at the table. The first lesson can begin the next time you are served a spinach-and-raw-mushroom salad. Looking at a slice through the middle of

What's in a Name? In scientific terminology, a mushroom is the fruiting body of a fungus. Everyday language, however, tends to be more vague. Some people would argue that the word *mushroom* should be limited to fungi with a stem and a cap, while others say the term includes all large and distinctive specimens, even if they have the shape of corals, brackets, or balls. There are no rules as to where to draw the line, but my preference is to use *mushroom* as an all-inclusive term to denote large, readily visible fungi. Molds and yeasts are certainly not "mushrooms."

Among the mushrooms there is a great variety of sizes and shapes, and it is these characteristics that will help the beginning mushroomer categorize specimens. The major types of mushrooms are as follows:

Category (common name)	General characteristics	Examples
Agarics (gilled mushrooms)	Cap and stem, with "gills" or partitions on the underside of the cap	White button mushroom
Boletes (sponge mushrooms)	Cap and stem, with tubes on the underside of the cap	King bolete (cep, porcini, Steinpilz)
Hydnums (toothed mushrooms)	Cap and stem, with "teeth" hanging from the underside of the cap	Hedgehog mushroom, bear's head

continued

Category (common name)	General characteristics	Examples
Clavarias (coral mushrooms)	Usually branched; found growing on the ground, looking like marine corals	Yellow coral
Tremellales (jelly fungi)	Irregularly shaped, often gelatinous and translucent; found growing on trees	Wood ears, cloud ears
Gasteromycetes (puffballs)	Generally spherical, growing mainly on the ground, sometimes on trees	Common puffballs, giant puffballs, earth stars, bird's-nests, stinkhorns
Ascomycetes (various types)	A very large group of fungi with many shapes and forms	Cup fungi (e.g, morels); truffles; ergot; most molds and yeasts

You'll notice that there is no category for "toadstools." Many people would say that "mushrooms" are edible and "toadstools" are not, and others would call all cap-and-stem mushrooms "toadstools" (after their shape). But *toadstool* is slang and is used to mean different things by different people.

Properly speaking, then, it is acceptable to call any mushroom a "fungus," but not all fungi are mushrooms. The Germans use the general term *Pilz* for all forms of fungus, macroscopic and microscopic, and the French call everything a *champignon*. An American friend of mine living in Paris was diagnosed by her local physician as having an oral *Candida* yeast infection. What the doctor actually said to her, in his best English, was: "Madame, you have a mushroom in your mouth!"

the mushroom, you will see the gills—most likely pink or light brown—beneath the cap, and the remnant of a membrane that covers them. This membrane, which is found in many but by no means all mushrooms, serves to protect the delicate spore-bearing surface from insects, raindrops, bits of dirt, and the like. Eventually, as the mushroom matures and the cap expands, the membrane breaks from the edge of the cap and becomes a ring around the stem. Timing is important: the membrane should break about when spores begin to be produced. Thus, the membrane plays its protective role until it is no longer needed.

A mushroom, then, is the end result of complicated biological phenomena. The reason the fungus goes to all this trouble is the same as the reason apple trees make apples or roses bloom—to produce seeds (or, in the case of the mushrooms, spores) for the propagation of the species. Reproduction by this method is advantageous for two reasons. One is that seeds or spores can be carried over long distances, allowing the species to be dispersed over large areas instead of remaining restricted to a single site. The other reason is that both spores and seeds are formed as the result of a sexual process, which allows the individual to interchange its genes with those of another individual. The consequence is that the species can evolve more rapidly and thus adapt efficiently to changing circumstances. Mushrooms provide good examples of the lengths to which living things go to ensure the survival of future generations.

3

UMBRELLAS AND
OTHER VARIATIONS

"The simplest and most lumpish fungus has a peculiar interest to us, compared with a mere mass of earth, because it is so obviously organic and related to ourselves, however remote. It is the expression of an idea; growth according to a law; matter not dormant, not raw, but inspired, appropriated by spirit. If I take up a handful of earth, however separately interesting the particles may be, their relation to one another appears to be that of juxtaposition generally. I might have thrown them together thus. But the humblest fungus betrays a life akin to our own. It is a successful poem in its kind. There is suggested something superior to any particle of matter, in the idea or mind which uses and arranges the particles."

So wrote Henry David Thoreau in his journal on October 10, 1858. By attributing to the "humblest fungus" a life akin to our own, he reminds us of the seemingly infinite ways that living things find to adapt to their environment. Mushrooms, for example, appear after a good rain, but they do not enjoy the actual downpour any more than we do. The moisture brought by the rain is required for their development, but mushrooms take steps to protect themselves from falling raindrops. They must do so because the spore-bearing surface, be it gills or tubes in the underneath of the cap, would be damaged by the impact of water hitting it directly. In other words, the familiar "capped" mushroom is umbrella-shaped for the same reason that we humans have invented umbrellas—to keep the rain out.

The *raison d'être* of mushrooms is to facilitate the spread of spores over as wide an area as possible. The microscopic spores of most fungi are light enough to be borne aloft by even mild air currents, a mode of dispersal that is generally more efficient than dispersal by water: as long as they are dry, spores may be scattered to distant habitats, whereas if they were washed away by water they would be more likely to disseminate over a smaller area. Conveniently, mushroom spores are quite resistant to desiccation and survive long trips aloft.

All kinds of strategies have evolved to maximize the dispersal of spores. I refer here not only to the umbrella shape of those fungi usually associated with the term *mushroom*. Fungi may take a great many different shapes—there are puffballs, truffles, coral fungi, or jelly fungi. In terms of numbers, the output of spores can be prodigious. A middle-sized mushroom with a four-inch cap may produce on the order of 20 billion spores over a period of four to six days, at a rate of some 100 million per hour. Giant puffballs may yield 20 trillion spores, a figure so large that it can be grasped only by comparison to the national debt of the United States. The most likely reason for the astounding number of spores is that most of them will not "make it"; that is, most will not germinate to become another mushroom. Some will land in an unfavorable environment, some will be crowded out by intense competition from other organisms.

Consider two common umbrella-shaped mushrooms, the agarics and the boletes. The gills of agarics and the tubes of boletes exist for two reasons. One is functional, to maximize the area of the surface on which spores are made, thus making it possible to produce them in larger numbers. The second reason is architectural: these structures are needed to help hold up the cap of the mushroom. Gills act much like the stays of an umbrella or like the "ribs" of a Gothic fan vault (see Figure 3.1).

The total surface area of the gills of a middle-sized specimen, such as a white button mushroom just an inch and a half in diameter, corresponds to that of a flat disk some 10 inches in diameter. The actual figure varies with the size of the specimen and the number and dimensions of the gills. No space is wasted, as you can determine for yourself by looking at the underside of a mature button mushroom. You will see that between the full-sized gills are smaller, incomplete gills occupying the extra spaces near the edge of the cap. The point is that having gills increases the spore-bearing surface by a very large factor. The same is true for mushrooms with tubes or tooth-like projections dangling from the caps in place of gills.

How are the spores set free from the cap? In fresh specimens at least, the surface of the gills or pores is moist and spores would cling to the mushroom, which would be of no help in their dispersal. Obviously, there would be little point to making a myriad of spores if they all remained close to the mushroom. But that doesn't happen, because spores are *forcibly ejected* from the cells that produce them. This is readily demonstrated by cutting the stem off the cap of a mature specimen and placing the cap on a piece of paper. In time, the ejected spores will make a "spore print," visible mounds of spores corresponding to the spaces between the gills or within the pores (more about spore prints in Chapter 4).

In most of the mushrooms, spores are made by specialized cells that line the surface of gills and tubes, known as the *basidia*. Each basidium makes two, four, or more spores. The release of spores has intrigued researchers for some time, but the precise mechanism is still not known. Several impelling forces have been invoked: a pressurized gas bubble, liquid under pressure, and some kind of electrostatic repulsion. A current suggestion, which seems as good as any, is that spores are released by a "whiplash" mechanism: When a spore matures, it remains attached to its mother cell by a slender stalk. A drop of liquid emerges near the base of the stalk and soon adheres to the parental cell, bending the stalk.

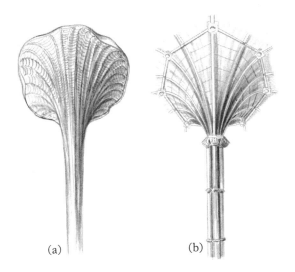

(a)   (b)

3.1
A case of "convergent evolution"? Mushroom gills and gothic vaults share interesting similarities in structure.

This puts the stalk under tension, "cocking" it, as it were. When the tension becomes too great, the stalk springs back, and the spore is detached and is "whiplashed" into space. Whether or not this proposed mechanism is correct, the propelling force is considerable: spores are ejected vigorously and travel in the air for almost two-hundredths of an inch. This may not sound like a great distance, but it represents 10 to 20 spore diameters, which is ample for the purpose of clearing the surface of the gill.

The distance that the spores travel must be just right. If the path traveled is too long, they could land on the opposite gill. Several factors minimize the chance of such mishaps.

One is that each spore is so small that, after traveling for about one-thousandth of a second over a distance of less than one-tenth of a millimeter, it abruptly loses momentum and falls straight down. In other words, spores do not travel through the air as a baseball does, approaching the ground at an ever-decreasing arc. Instead, the trajectory of mushroom spores begins horizontal to the ground, but then it takes a ninety-degree turn into a sharp vertical drop. This is nothing short of the ultimate curve ball that every baseball pitcher dreams of. Because of the propulsive force of their release, spores made at the top of the gill have the same chance of falling free as those made at the bottom.

Mushrooms have other geometric arrangements that help their spores fall freely. As the specimen matures, the cap expands and the space between the gills increases, giving spores an even greater chance of clearing the adjacent gill. In many species, the cap keeps expanding with age and may become concave, with the consequence that older specimens have a funnel shape. In addition, mushrooms exhibit geot-

3.2
An *Amanita* rights its cap after having been placed on a table sideways for ten minutes.

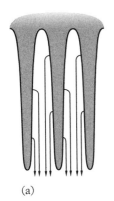
(a)

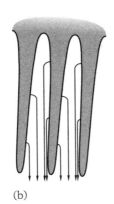
(b)

(c)

3.3
Wedge-shaped gills
allow spores to fall
free even if the cap
is slightly tilted.

ropism; that is, they know which way is up, and they keep their caps upright and the gills as vertical as possible. Certain species will even right their caps when placed on a surface sideways—*after* being picked (see Figure 3.2). The amanitas are champions at this game. After not much more than thirty minutes, the stem of a "toppled" amanita will have bent, and its cap will stand upright.

The shape of the gills, as well as the shape of the cap, promotes spore dispersal. In most gilled mushrooms, the gills are wedge-shaped in cross section, broadest near the cap and narrowest at the margin. This allows the spores to clear the surface even if the gills are slightly tilted. (See Figure 3.3.)

Spore dispersal is an example of conspicuous production. In this case, lavishness is necessary: rare is the spore that germinates into triumphant fungal growth and helps the species spread in the environment. Such "wastefulness" is not unlike the production of millions of unproductive sperm cells in the human male. There is lavishness also in the variety of forms in which mushrooms serve as dispersers. Standing apart from the other gilled mushrooms are the shaggy manes and other inky caps *(Coprinus)*. What is unique about these species is that after the spores are liberated, the now barren gills do not remain intact, as they do in other mushrooms. Rather, they become liquefied and fall to the ground as drops of dark, inky liquid. The liquid produced is sufficiently ink-like that it was used for writing in colonial times. In these fungi, spores are produced in waves, starting at the gill's edge and then moving upward, whereas in most other gilled mushrooms spores are made randomly all over the gill. The used-up portion of the *Coprinus* gills, located beneath the region where spores are being made, is no longer in the way of spore dispersal. Not surprisingly, *Coprinus* gills are not wedge-shaped but instead have parallel sides.

The *Coprinus* way of life does, however, pose a different problem. Some, like the mica caps *(C. micaceus)*, grow in extremely tight bunches of hundreds of specimens, which means the caps are unable to expand. How do the gills of the tightly packed mushrooms keep from collapsing upon one another? To keep this from happening, coprini mushrooms produce gigantic sterile cells that protrude from the gill surface, the *cystidia*. These cells serve as props or struts, keeping the gills from sticking on one another and maintaining the air space between them. (See Figure 3.4.)

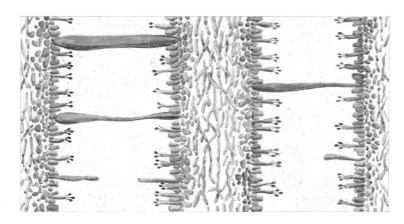

3.4
A section through an
inky cap reveals the
presence of long, strut-
like cells (cystidia) that
span the space between
adjacent gills and keep
them from collapsing
on one another.

This neat piece of structural engineering allows efficient
spore dispersal despite the tight clumping of the mushrooms.
Coprinus mushrooms have not sacrificed efficiency for to-
getherness.

Fungi that are not umbrella-shaped solve the challenge of
spore dispersal in other ways. The coral mushrooms, so
called because they form upright branches, have a spore-
bearing surface that is unprotected and more or less vertical.
These mushrooms spread their spores by a combination of
two mechanisms: when the weather is dry, dispersal is by
air; when it rains, via water. Corals do not have as large a
spore-bearing surface as cap-and-stem mushrooms, but they
have the advantage of not having to construct more complex
structures at great energy cost. Their method of spore dispersal seems a simpler mechanism and, fittingly, the coral

fungi are considered by some researchers to be primitive on the evolutionary scale.

Spore-dispersal strategies reach the height of ingenuity with the puffballs and their allies, the Gasteromycetes or "stomach fungi." The simplest and most passive of these mechanisms is the one used by the larger puffballs. When they mature, the surface layer of the puffball disintegrates and the dry spores are blown away by the winds. Eventually, little is left except for the cottony base of the puffball.

Unlike their big cousins, the common little puffballs seen both in fields and forests retain their surface coverings for a long time. One can find nearly empty puffball bags months after the fruiting season is past, sometimes even a year later. If you have a chance to examine one of these dried-out, misshapen brown pouches, you will see that the spores must have escaped via the hole at the top. To be convinced, all you have to do is to squeeze the pouch, and the remaining spores will "puff" through the hole. This aperture, called the *ostiole,* or "little door," is formed about the time the spores mature and the insides of the puffball have dried out, long after the specimens have lost their gastronomic appeal. Under normal circumstances, without the help of a human puffball-squeezer, how do the spores escape through this small hole? An English mycologist, P. H. Gregory, figured out in 1949 that they exit with the aid of raindrops. Gregory used a special camera to record the events and found out that when raindrops fall anywhere on the puffball surface, the leathery outer skin is momentarily depressed and the spores are "blown out" the ostiole. In other words, the surface covering acts as a bellows. These puffballs not only do not mind rain, they actually depend on raindrops for spreading their spores. Gregory calculated that in a fairly rainy place, as in much of England, a puffball gets hit by as many as a quarter of a million raindrops in its lifetime.

Even more complex strategies to ensure spore dispersal have developed in an odd-shaped group of mushrooms known as the earth stars. It would be fair to call them the construction engineers of the fungal world. Earth stars are puffballs surrounded by "rays" that give them a star-like appearance. The rays are formed from an extra covering that splits into sections from the top down, much as an orange might be peeled. As the rays bend outward, they push away any leaf litter that may have covered them and that would have kept raindrops from hitting the surface of the spore sac. Some earth stars (e.g., *Geastrum fornicatum*) even raise the spore sac on a pedestal, as if reaching up for the rain!

In sandy soils and other arid areas one can find a kind of earth star that puts up an unusual show. The rays of the hygrometer earth star (*Astraeus hygrometricus*) open up when wet, but they refold to cover the puffball when dry. This mushroom is very accommodating—one can force the covering to switch back and forth between the two forms a dozen times just by letting these mushrooms dry and then wetting them again. Children and adults alike will enjoy the show. This outside covering is not just for show, of course, or for weather prediction; it exists to maximize spore dispersal. These specimens are easily detached from their underground filaments, and in the round, folded-over form they can be pushed along by the wind, spreading their contents of spores. Think of them as rolling salt shakers. When it rains, the rays open up and the puffballs may become upright and "puff" in the usual manner.

Other startling feats of fungal engineering are seen among the bird's-nest fungi. Members of this group are often overlooked by mushroom hunters since they are small, a quarter inch or less across. They grow on fallen sticks and other vegetable debris and can best be seen when growing in bunches. The specimens are shaped like little urns that, in the young, are covered by a protective lid. When the spores are made, the lid is shed, revealing within each cup a collection of a half-dozen or so round, hard sacs that look for all the world like miniature bird's eggs (see Figure 3.5). Actually, each sac, called a *peridiole,* contains a multitude of spores.

How do the spores escape from the peridioles? The answer, researched thoroughly by the Canadian mycologist H. J. Brodie in the 1950s, has to do with the shape of the cup. When a raindrop falls into a bird's-nest cup, it hits with such force that one or more of the peridioles are ejected. The design of the "splash cups" is so efficient that when the peridioles pop out, they travel several feet from their "nest." And that is not all. As the peridioles fly into space, they trail a long thin thread that is very sticky. When this thread touches the stem of a plant or a leaf, it clings to it, keeping

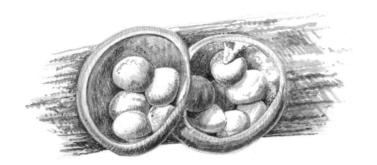

3.5
The sacs of spores
inside the "splash cup"
account for the common
name of these mush-
rooms, the "bird's-nest
fungi."

the peridioles suspended in the air. Why do they do this? Brodie proposes that dispersal of bird's-nest spores may be carried out by animals that accidentally eat the peridioles along with vegetation. The spores pass unharmed through the animal's digestive tract and are deposited elsewhere. Quite a few of these fungi are coprophilic (feed on dung), and thus find their most suitable habitat as soon as they are shed from the animal. Thus, the intricate body plan of the bird's-nest fungi may have evolved so that herbivorous animals will unwittingly swallow peridioles and deposit the spores at distant sites.

Occasionally, the easily overlooked bird's-nest fungi make their presence felt. A friend of mine once found that his car was spattered by little brown specks. Puzzled, he recalled the "yellow rain" reported to have fallen in Vietnam during the war, which was believed by some scientists to consist of bee droppings. Possibly these specks might be made up of insect feces. When my friend looked carefully, though, he could find the specks on one side of his car only. He realized that the car was parked next to an area of his garden that was covered with mulch and, when he bent over to inspect it, he found that it was populated with hundreds of tiny bird's-nest cups. The rain had splashed up a large number of peridioles, many of which had landed on the car surface. A very modern form of spore dispersal, making use of automobiles!

Why do mushrooms make so many spores and go to such lengths to make sure they are dispersed away from the site of their birth? Most spores won't succeed in the wild—they will make landfall in an inappropriate habitat, or get eaten by small invertebrates. Even if the spores go on to germinate, the emerging filaments may not thrive in competition with other living forms already established at the site. Thus, the success of a given species is enhanced by dispersing a large number of airborne spores over a large area, sometimes miles away from their origin. Survival of the fittest requires that a species not only out-eat and out-fight its neighbors, but also the ability to spread into new territories. The farther a species travels, the more likely it is to find new sites for growth and development. Nature is not just "red in tooth and claw" —it is filled with wanderlust as well. The mushrooms are a fine example of this strategy.

Niji kiete ato wo tazuneru kinoko kana

The rainbow faded,
And after that, seeking
Mushrooms.

Yusho

COLLECTING, IN SOLITUDE AND IN GROUPS

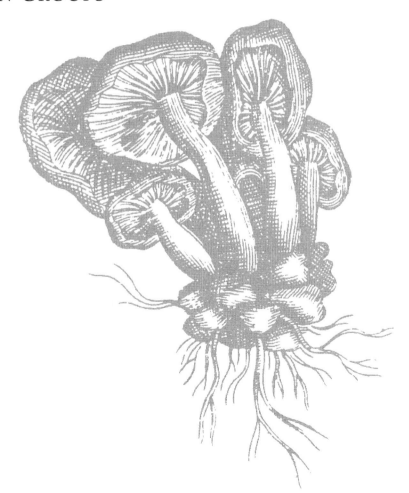

4

A FORAY IN THE WOODS

"Bolé, bolé!" the children called out whenever they found a mushroom. We were in the midst of a mushroom hunt in the mountains above Barcelona, where I had been taken by my friend Josep, his wife, and their three- and four-year-old boys. The woods were full of specimens and the children were enjoying the hunt. We had driven for less than two hours to the woods of Montseny. Deciduous trees were mixed with evergreens, the terrain was rocky, and there was practically no undergrowth, giving the woods a manicured look. There was nothing primeval about this forest. Mushrooms had popped up from the bare ground, which made spotting them easy and certainly less of a challenge than what I am used to in the denser and less groomed forests in New England. We found about forty different species of mushrooms in less than an hour, including a number of old friends and some that looked similar to the kinds I know, yet sufficiently different that they could be other species. Among the familiar ones was the sort that is everyone's idea of a mushroom, the classic fly agaric, *Amanita muscaria,* with a crimson cap adorned with white patches. Josep was happy to be able to show me the mycological richness of the area, but he did not seem particularly surprised at our success because he had been mushrooming in this bountiful area since childhood, as had his parents and grandparents.

When the children yelled "bolé!" they were using an ancient term for a mushroom derived from Latin, which suggests both the independence of the Catalan language and the uninterrupted connection of the Catalans with ancient Rome. *Bolé* comes from *boletus,* the name the Romans used for any mushroom with lamellae or gills on the underside of the cap. Linnaeus adopted a completely different meaning of the word nearly two hundred years ago. Outside of Catalonia and a few other places, *bolete* now means a mushroom with tubes instead of gills underneath the cap. In Roman times, mushrooms with tubes had the quaint name of *suilli,* "little pigs," hence the Italian *porcini.* The children's cries made me think of how long these words had been uttered in that part of the world.

When we returned to the car, Josep explained to his children what we had found, using Catalan names for the specimens in the baskets. He selected an orange-colored mushroom with green spots and, using his pocket knife, made several cuts across the gills. Immediately, a rich orange liquid oozed out and turned green within a short time. He explained that this was one of the milk caps, a *Lactarius.* This particular specimen belonged to the species *Lactarius deli-*

ciosus, or, in Catalan, *rovelló.* This species had been prized since antiquity, as witnessed by the fact that it or one of its close allies is depicted in a fresco unearthed in Pompeii. The painting shows a group of eight fresh-looking mushrooms with a duck ready for the oven. This 1,900-year-old fresco may be the oldest depiction of mushrooms known, at least in the western world.

Our harvest that afternoon seemed bountiful enough, but I later realized that we had not even scratched the surface, as it were. As we were starting to drive away, two men came out of the woods, laden with huge plastic bags, one full of porcini *(Boletus edulis),* the other with *rovellós.* Both loads were destined for market and, sure enough, I saw these and other wild mushrooms offered by the boxful in Barcelona's big market adjacent to the Rambla de Catalunya, the city's main boulevard.

Possibly the most popular outdoor activity that is compared to mushroom hunting is bird watching. The two share several features: they both call for a love of the outdoors, considerable zeal, and the ability to put up with frustration. In my mind, however, there are advantages to mushrooming. The specimens don't fly away; you don't have to be quiet in the woods; and you don't have to worry about aerial bombardment. On the other hand, starting out on the path of identification is harder for the mushroomer than for the birder, one reason being that one begins with a more slender base of reference. Every fledgling bird watcher knows that sparrows are songbirds, sea gulls shorebirds, and hawks raptors. Even the beginner can usually tell what large group a new bird belongs to. Most people are more limited when it comes to mushrooms and may only be able to discern that a specimen is of the cap-and-stem variety, a puffball, or a bracket fungus. This is useful as a first step, but a more extensive framework is needed.

Some mushrooms are large and readily advertise their presence. Others are small and unassuming and reveal themselves only to those with patience and keen eyesight. I feel especially attracted to some of the inconspicuous kinds, perhaps because finding them is a small triumph. After a good autumn rain, the bark of the English elms on my street in Newton, a suburb of Boston, sported the inconspicuous *Mycena corticola*. Although small brown caps appear by the dozens, they were still easy to miss. They are most noticeable when they grow on contrasting patches of shiny green moss. Amateur mycologists lump small mushrooms that are hard to identify into a group they call LBMs, for "little brown mushrooms." The *Mycena* on the bark should then be called VLBMs, "very little brown mushrooms." They are less than a quarter-inch across and have a dimple in the middle of their nicely rounded caps. Turned upside down, they reveal a few well-spaced blades (the gills). The thin stems stick straight out of the bark and elegantly curve upward to hold the caps erect. You may find them next fall as long as you don't mind having the neighbors see you standing in the rain staring at the side of a wet tree. These mushrooms are not limited to elms but are seen on many of the larger deciduous trees. "Edibility unknown," various field guides say. No matter, it would take a gross to make a small mouthful.

The most common mushrooms found in the North American forests are the agarics and boletes. These can readily be identified by turning a specimen over and looking at the underside of the cap. Agarics have gills whereas boletes have tubes tightly packed together so that the undersurface of the cap appears to be full of holes and resembles a sponge. Gills and tubes, as I explained in Chapter 3, are the surfaces on which the spores are made.

One finds mushrooms most predictably in forests and, if there has been a lot of rain, in open spaces, including suburban lawns. The woods are the better bet since they retain moisture for longer periods; in North America, that is where mushroom collectors head most of the time. In places that tend to be more humid and that have been deforested long ago, such as England, "open-space mycology" is more widely practiced.

Each region of the globe has its own repertoire of wild mushrooms, depending on the prevailing vegetation and weather. Because my own experience is based largely in New England, I find it humbling to pick in California, Oregon, or Washington, where I cannot yet identify many of the locally plentiful species. How much overlap is there? About half of the mushrooms of eastern North America are also found in the West. The differences are even greater between the temperate zone and the tropics, and there is relatively little overlap in the species found in these regions of the world. There are exceptions, though; certain mushrooms are nearly cosmopolitan. Morels, for example, are as well known and prized in India and much of the rest of Asia as they are in Europe and North America.

At times one finds certain kinds of mushrooms where one least expects them. Often, foreign species are unwittingly introduced into a new area by association with cultivated trees imported from elsewhere. Many forest-dwelling mushrooms are "mycorrhizal," meaning that they live in symbiotic association with the roots of trees (see Chapter 12) and are brought along with transplanted seedlings. I was surprised, for instance, to find an edible mushroom known as Slippery Jack (*Suillus luteus*) in the Cotopaxi National Park in the highlands of Ecuador. This species is relatively common in Europe and eastern North America but is not known to grow in the Andes. The ones I found, however, were growing in a huge plantation of introduced Monterey pines (*Pinus radiata*). Not only was this encounter unexpected, the harvest was exceptionally bountiful. Because the growing season at this location near the equator is almost year-round, the growth of this species may constitute one of the largest yields of mushrooms per acre anywhere.

Mushrooms can also be found in unexpected habitats. Surprisingly, they inhabit deserts and other arid areas, although certainly not in abundance. On Cape Cod and other shores, they can be found between the sand dunes and terra firma, where the area is frankly sandy but has some solidly implanted vegetation. In this habitat, grasses and shrubs reach sizable proportions but leave patches of naked sand between them. It is in these places that fungal life can be found in late summer. The most mushroomy-looking specimen here is called *Laccaria trullisata,* a species with a cap that sticks out an inch or so from the sand surface but which has a tuberous stem reassuringly buried in the sand to a depth of four or five inches. Happy as a clam.

Identifying wild mushrooms is a chancy proposition— some can readily be told apart, others need detailed study, perhaps even examination with a microscope. Like Rumpelstiltskin, who would not reveal his name, some species seem to conceal their identity almost willfully. With a little experience, however, a hunter can place most of the local mushrooms into one of a few large taxonomic groups. I reckon that about 80 percent of the mushrooms found by casual collectors in a given area belong to some 200 species, a large but not unmanageable number of varieties to learn. All told, there are thousands of species and many have not yet been properly described. The number of all fungal species on record is about 69,000, but most of these are molds and yeasts, not mushrooms. The eminent British mycologist David Hawksworth calculated that this is a gross underestimate and that the actual number of fungal species may reach 1.6 million. Since nobody knows how many species of mushrooms there are on earth, or even in any particular region, there is room for a lot of work. In the early 1970s, at a mushroom foray in New Hampshire, I had the experience of

	Common name	Species name
Fifty North American Mushrooms	Artist's fungus	*Ganoderma applanatum*
	Beefsteak fungus	*Fistulina hepatica*
	Blewit	*Clitocybe nuda*
	Blusher	*Amanita rubescens*
	Bird's-nest fungus	*Cyathus, Crucibulum* species
	Brick cap	*Naematoloma sublateritium*
	Caesar's mushroom	*Amanita hemibapha* (in U.S.)
	Cauliflower mushroom	*Sparassis crispa*
	Chanterelle	*Cantherellus cibarius*
	Chicken mushroom	*Laetiporus sulphureus*
	Coccora	*Amanita calyptrata*
	Death cup	*Amanita phalloides*
	Destroying angel	*Amanita virosa, A. verna, A. ocreata*
	Dog stinkhorn	*Mutinus caninus, M. elegans*
	Earth ball	*Scleroderma citrinum*
	Earth stars	*Geastrum, Astraeus* species
	Enoki	*Flammulina velutipes*
	Fairy ring mushroom	*Marasmius oreades*
	False morel	*Gyromitra esculenta*
	Fly agaric	*Amanita muscaria*
	Giant puffball	*Calvatia gigantea*
	Hedgehog mushroom	*Hydnum repandum*
	Hen-of-the-woods	*Grifola frondosa*
	Honey mushroom	*Armillaria mellea*
	Horn of plenty	*Craterellus cornucopioides, C. fallax*

continued

Common name	Species name
Horse mushroom	*Agaricus arvensis*
Inky cap	*Coprinus* species
Jack O'Lantern	*Omphalotus illudens*
King bolete	*Boletus edulis*
Lobster mushroom	*Hypomyces lactifluorum*
Milk caps	*Lactarius* species
Matsutake	*Armillaria ponderosa*
Meadow mushroom	*Agaricus campestris*
Morel	*Morchella* species
Old Man of the Woods	*Strobilomyces floccopus*
Orange peel fungus	*Aleuria aurantia*
Oyster mushroom	*Pleurotus ostreatus*
Parasol mushroom	*Lepiota* species
Porcino (usually the plural, Porcini)	*Boletus edulis*
Prince mushroom	*Agaricus augustus*
Shaggy mane	*Coprinus comatus*
Shiitake	*Lentinus edodes*
Slippery Jack	*Suillus luteus*
Stinkhorns	*Phallus, Dictyophora* species
Sulfur shelf	*Laetiporus sulphureus*
Sulfur tuft	*Naematoloma fasciculare*
Turkey tails	*Trametes versicolor*
Wine cap	*Stropharia rugoso-annulata*
Witch's butter	*Tremella mesenterica*
Yuck-on-a-stick	*Exidia glandulosa*

showing a new (to me) specimen to a great authority on American mushroom taxonomy, the late Alexander Smith. His books on mycology take up about three feet of bookshelf, so I was confident that he knew everything there was to know and would easily identify my specimen. Peering over his glasses, he looked at it carefully, turned it over a few times, wrinkled his nose, and finally uttered: "There is no such thing!"

Not everyone needs to approach mushroom identification with equal fervor. For most people, a reasonable goal is to get to know several edible kinds with assurance. Learning to recognize the most frequent mushrooms in a given area requires work and dedication, but considerable familiarity can be acquired over a couple of collecting seasons. In time, one gets a sense of which mushroom species are common to a given area and which are not.

The most common types of mushrooms are listed in the accompanying table. This listing does not correspond to scientific taxonomic categories (in the language of systematic biology, they belong to different levels of "taxa"), but provides convenient "cubbyholes" for the mushroom amateur. If a dozen mushroomers went out on a good collecting day, they would come back with representatives of most of these kinds. The fastest way to learn about wild mushrooms, in my opinion, is to enlist the help of a trustworthy person as a "tutor" and to augment your field experiences by studying one of the many available field guides (a list is provided at the end of the book, in "Resources"). The reason you need both and cannot rely solely on personal advice is that popular knowledge of mushrooms does not travel well. Familiarity with mushrooms of one region of the world may be misleading in another. Certain species found in certain areas may be similar but not identical to the ones that are found elsewhere. There is strong temptation for people with experience elsewhere to "recognize" local specimens as familiar ones. Here is an example: In the late 1980s and again in 1996, a number of Laotians living in California and Oregon came down with extremely serious cases of food poisoning from eating one of the most lethal of our mushrooms, *Amanita virosa,* the "destroying angel." They had mistaken this poisonous mushroom for a similar-looking species that in Southeast Asia is one of the great edibles, much prized and frequently sold in the markets.

All one's senses come into play when trying to identify wild mushrooms. The eyes reveal the most, especially when they are trained to recognize distinguishing characteristics, such as shapes, sizes, and colors of the cap and stem, not to mention special attributes such as patches on top of the cap or a ring around the stem. As is true for tasting wines or cheeses, the sense of smell can be very useful. Many mushrooms have characteristic odors, such as the scent of fresh corn, coconut, anise, almond, garlic, or plain "mushroom." There is indeed a "mushroomy" smell. It is the attractive aroma typical of the white buttons in stores, and it is found in a large number of wild mushroom species as well. The main chemical involved has been identified as 1-octen-3-ol,

4.1
The chemical
responsible for
the "mushroom"
flavor of many
species is
1-octen-3-ol.

$$CH_3-(CH_2)_4-\overset{\displaystyle H}{\underset{\displaystyle OH}{C}}-CH-CH_2$$

which has been isolated from many kinds of mushrooms (see Figure 4.1). At the other end of the olfactory spectrum are the stinkhorns, which in their mature form emit an odor not unlike that of seasoned carrion.

The sense of touch comes in handy as well, because mushrooms often have distinctive surface textures. Some have a velvety feel, others are tough or even brittle. Beginners tend to squeeze mushrooms hard, as if they want to force the truth out of them, but mushrooms do not respond well to this treatment. Much is revealed by tasting a small piece of a mushroom and spitting it out. The taste of most mushrooms is mild, but some are characteristically pungent or bitter. The sense of hearing is less likely to give clues as to the identity of a mushroom, but the violent discharge of spores in some fungi can actually be heard by placing the specimen near one's ear!

I beg the reader to accept the fact that there are no general rules for telling an edible mushroom from a poisonous one. The myth that a poisonous mushroom will turn silver coins or spoons black *is* just a myth—the most lethal mushrooms won't pass the test. Short of a sophisticated chemical or biological analysis, there is no way to tell. My friend Larry Millman claims that his pet cockroaches will eat edible amanitas and disdain the poisonous ones. Even if this proves to be true, how many people have a colony of hungry cockroaches at hand?

"There is no royal road to the knowledge of Fungi," said the Reverend John Stevenson over a hundred years ago. It takes some effort and experience to know with certainty the identity of a mushroom. I have avoided being poisoned by strictly following two obvious rules. One is to proceed with caution. When I try a new mushroom, I first sample a very small piece and keep the rest in the refrigerator. I will eat it the next day if there has been no indication of unwholesomeness. The other rule is to make positive identification of everything I eat. It's not enough for me to know that a specimen is not one of the *known* poisonous mushrooms because there are lots of species for which reliable information on edibility is lacking.

Truly, there is no royal road to knowledge, or safety: you should know exactly what you are going to eat.

When I lived in Boston, I went a-mushrooming every chance I had. My usual haunts were the nearby woods, usually less than a half-hour from my house in the suburb of Newton. I could expect to find a nice variety of mushrooms in August or September, and as long as there had been a reasonable amount of rain recently. In the Northeast, the ideal conditions for mushroom picking—frequent, soil-soaking rainfalls—do not occur every year in the late summer or fall, so I tried to take advantage of every good opportunity.

The equipment for my forays, then and now, is a stout basket about sixteen inches long, a Bowie-type knife, a magnifying lens (which I wear on a string around my neck to make sure I don't lose it), insect repellent, a notebook, a roll of wax paper, and a whistle. I always carry a whistle for reassurance because with all the looking down I do, I tend to get lost. Actually, I am not sure what whistling in the woods would do for me, since I usually don't make arrangements with others to respond to these audible signals. A compass would probably be useful, but I don't want to add orienteering to all the things I should already know. The wax paper is for wrapping specimens, each into a neat individual package with twists at the ends, which keeps them from getting squashed and soiled in the basket.

My usual haunts used to be the common type of forests of New England, the deciduous woods mainly of oak and maple. If rain had been sparse and the weather hot, I would stay away from pine, spruce, or hemlock woods because coniferous forests do not hold the moisture as well as deciduous ones. On the other hand, when conditions are right, evergreen woods display stupendous growths of mushrooms, which are much more visible here than in deciduous forests because there is less undergrowth. In open-ground woods, it is sometimes possible to see mushroom upon mushroom as far as the eye can wander.

Deciduous forests have considerable undergrowth, so I found it best to stick to the paths to avoid catbriers and other thorny shrubs. It is not necessary to mortify the flesh to find

mushrooms, many of which prefer slightly open spaces any-
how. It is also rewarding to look for mushrooms on dead
standing or fallen trees, especially late in the season, when
ground-inhabiting species become scarce. How lucky I
would be depended on the weather and the terrain, but on
a good day, mushrooms seem to be everywhere.

By the time I walked for a hundred feet or so into a forest,
I usually had some idea what the pickings were likely to be.
Would this be a good day, or would I have to scrounge
around to find just a few specimens? It is not always possible
to tell, because at times you might find lots of mushrooms
on one side of a hill and few on the other, depending on the
wind pattern and how well the terrain holds the moisture.
Or, in what seems to be premeditated malice, sometimes you
will find some gorgeous specimens at the beginning of your
walk and little else thereafter. However, I believe that one
develops a sense for the abundance of mushrooms and, in
time, for the side of the hill that is best for picking.

How different is the collecting experience in the woods
near my new home in southern California! On a casual walk
in the live oak forests near San Diego, one very seldom sees
a mushroom popping out of the ground. Happily, this dearth
of specimens is totally misleading, because during the season mushrooms are abundant, although they remain hidden
from sight. What's going on? The leaves of the live oaks are
leathery and tough, and with the scant rainfall they decom-
pose very slowly. Here, the ground is covered by a thick
protective layer, sometimes four to six inches deep. Under-

neath, the moisture is retained for a long time, and even the occasional rains go a long way. Even robust mushrooms, however, have difficulty emerging from this thick layer of duff. Instead, they push the leaves up into visible mounds. (This also happens in the Northeast and the Midwest for the short period after the leaves of deciduous trees have fallen but before they are tamped down by the next rains.) So, the mushroom hunter in southern California is really a "mound hunter." Each mound holds a surprise, as it's impossible to know ahead of time what it holds. Mushrooms in this area seem bashful, as though they like to hide from view, qualities not usually associated with Californians.

Let us now pretend that you, my reader, and I are going on a mushroom walk. Were this your first such walk, I would suggest a few do's and don'ts. If you want a mushroom for the purpose of study and identification, you should collect the specimens whole, including the underground portion. If you want to eat what you pick, each mushroom should be cleaned as much as possible before being placed in the basket. Resist the urge to become a human vacuum cleaner and pick every specimen in sight. The fewest number of specimens should be gathered consistent with your purpose. Mushrooms are a renewable resource, but once they are picked they are no longer there for others to see and enjoy. If your intent is to study a specimen, it will suffice to take just a few individuals, especially if they represent different stages of development. Only fresh, inviting specimens should be picked. Do not practice necromycology! When the pickings are lean, some people collect old, decaying mushrooms, perhaps in vain hope. Not only do these not add anything at all to the kitchen, they are often confusing for someone attempting to make an identification. And, finally, it goes without saying that mushroom pickers should disturb the environment as little as possible and leave few or no traces of their activities.

On a good mushroom day and in a good picking area, it is not unusual to come upon thirty to fifty different kinds of mushrooms in an hour or two of walking through the woods. How should we try to identify them? Since we have to start somewhere, we might first (and arbitrarily) consider the most commonly found group, the gilled mushrooms. In any given area, gilled mushrooms—or agarics, as they are semi-technically called—comprise thousands of species. Nobody, not even the experts, can aspire to know them all, thus it makes little sense to try to identify every specimen one runs into. It is both possible and desirable, however, to learn how to place each one in its general taxonomic cubbyhole.

The gilled mushrooms are classified by the color of their spores, which ranges from white or other light colors to pink, light brown, dark brown, or even black. It takes a little work to determine the spore color, because you can't tell just by looking at the gills, which are sometimes of a different color. For instance, the common white buttons of the supermarket have pink gills but make dark brown spores (as you will see if you compare the young buttons with the more mature portobellos, which are becoming increasingly available in the market).

To find out the color of the spores, make a *spore print.* Take a fresh specimen, cut off the stem at the point of attachment to the cap, and place the cap on a piece of paper. Cover the cap with a glass or a cup to preserve moisture. In an hour or two, enough spores will have dropped to make a deposit on the paper that is visible to the naked eye. It works only as long as the specimen is fresh (dry caps shed few if any spores). Because deposits of white spores are not easy to see on white paper, use colored paper or make the spore print on a sheet of transparent plastic (such as a kitchen wrap) that can be held up against different backgrounds.

My next suggestion is to concentrate for a while on a particular genus. I recommend choosing the genus *Amanita,* the reason being that its members are distinctive and common. Prudence and a sense of self-preservation demand this approach, for the amanitas include the species responsible for most fatal poisonings. Not every amanita is deadly, and some are eminently edible, but everyone who picks mushrooms for eating purposes should be able to recognize them on sight. Another reason for starting with this genus is that its members have a number of visible characteristics that make it relatively easy to tell them apart from other kinds of mushroom. Plus, focusing on these characteristics teaches one a lot about mushroom morphology in general.

The amanitas are well described and illustrated in just about every field guide. The archetypal mushroom with a red cap and white spots, the fly agaric or *Amanita muscaria,* is an example. Distinguishing features of the amanitas include different-colored patches of tissue on the cap (although these come off easily and can be washed off by rain), a membranous ring hanging down from the stem in mature specimens, and a swollen base of the stem, sometimes encased in a membranous cup. In the New England woods, one type of amanitas has a cap of yellow-orange that fades gradually toward the edge. This beautiful, medium-sized mushroom, about four inches in height, is a common eastern species, *Amanita flavoconia.*

Amanita pickers should be mindful to collect the entire specimen. If the ground is soft, pick it up from the base of the stem with your fingers; otherwise, dig around it with a knife or trowel. It is important to include the underground portion because, as is true for many mushrooms, much information is obtained from looking at the bottom part of the stem. Early in their development, amanitas are covered by a membrane, which breaks up as the specimen matures. The upper portion of the membrane breaks up into the patches that are left on the cap, and the lower part becomes the cup-like structure at the base of the stem of some amanitas. With most species of this genus, however, the bottom part of the membrane collapses onto the bottom part of the stem and is hard to distinguish from the stem itself. In *A. flavoconia,* a common amanita in the Northeast, the swollen bulb at the base is covered with darker yellow bits of tissue. As with many wild mushrooms, this one does not have a generally accepted common name. In some books, it's called "yellow patches," in others "yellow wart." Authors of field guides make up common names with gusto and disagree with one another as freely as scientific taxonomists do. In reality, only about fifty common names are in general use throughout North America (see the list on pages 64–65).

In any given locality there may be several hundred different kinds of amanitas, but usually a dozen or fewer prevail. One kind reddens on bruising (*A. rubescens*), another is pure white and has a stem that emerges from a membranous, vase-like cup (*A. virosa,* one of the deadliest in North America). An unlikely find in eastern North America is *A. phalloides,* the species that causes most of the deaths in Europe and in the western United States, but it has occasionally has been reported in the East, where it also has caused fatalities. Late in the fall, one of the most common mushrooms in the eastern woods is an amanita with a pronounced smell of raw potatoes (*A. citrina*).

To continue with the white-spored gilled mushrooms, let us stick to the most common kinds first. Luckily, some, like the *Amanita,* are easy to identify, whereas others can become a demanding challenge. A group that is as abundant and nearly as easy to recognize as the *Amanita* is the *Russula.* Russulas lack the rich morphological details of the amanitas, but they can be readily identified by their brittle texture. When you snap the stem apart, it will feel like breaking a piece of chalk. An even more destructive russula test is to throw a specimen against the nearest tree trunk: russulas will shatter, other mushrooms will stay pretty much whole. Russulas are medium- to good-sized mushrooms, with caps two to five inches across and a stem from one-half to one inch wide. There are at least two hundred different russulas in the northeastern United States alone, making it difficult to identify the species even though it is relatively easy to recognize the genus.

Russulas are generally edible, but most are considered to be nearly tasteless. Some russulas are distinctly peppery, which readily distinguishes them from other members of the genus. Different species vary greatly in color, with different hues and intensities of red dominating (*russula* means "reddish" in Latin). Other species are white, yellow, green, orange, tan, or purple. Russulas are so common that they are to mushroom hunters what sparrows or starlings are to bird watchers. Although they are the "weeds" of the mushroom world, they can be appreciated for their beauty and for adding brilliant colors to the forest. Besides, if russulas are present, can other mushrooms be far behind?

There is one other group of brittle mushrooms, the milk caps or *Lactarius*. The mushroom I found in the forest near Barcelona, *Lactarius deliciosus,* belongs to this group. Milk caps are so called because milk-like latex flows out when a specimen is cut, which makes them easy to identify. The beginner should be grateful for such a readily recognizable property, truly God's gift to mycologists. One need only cut off a piece of the cap or slice the gill area and watch for a liquid to exude. In practice, as in many things mycological, this doesn't always work so well, because when milk caps get the least bit dry they yield little or no liquid. Besides, some species release only a little bit of latex even when fresh (a magnifying lens is occasionally useful), while others are intrinsically big producers. A drab-looking species, *L. lygniotus,* discharges so much "milk" that it easily qualifies for a mycological Holstein prize—a tablespoonful is not an unusual yield for a mature specimen.

Lactarius milk is highly varied in color and consistency. In some species, the liquid is watery and colorless, in others opaque, thick, and brightly colored. In some *Lactarius* the color of the milk is unvarying, in others it changes, sometimes within a few minutes, from white or whitish to bright yellow, green, or red. In yet other species, the milk oozes out colored from the beginning. One kind, *Lactarius indigo,* has milk of deep blue, which is an unusual color in mushrooms. As if to emphasize that they have nothing to do with mammalian milk, some milk caps produce a very peppery latex that burns the tongue. One wonders if, in appropriate amounts, hot *Lactarius* mushrooms may find a place in the kitchen along the myriad varieties of *Capsicum* peppers used in Indian and Mexican cookery ("Enchilada de Lactarius," anyone?).

Milk caps tend to have a symmetrical shape, often showing concentric rings on their caps, and have a dry surface that gives them a distinctive look. Recognizing a lactarius from six feet away, where you can't see the milk, is an intermediate-level mushrooming skill, similar to the ability to recognize a black-capped chickadee among the songbirds or the tufted loosestrife among the wildflowers. In my opinion, the champion edible lactarius is *L. deliciosus,* although some people do not believe it deserves the name. Perhaps this species varies sufficiently from region to region to explain the differences in opinion. I find that *L. deliciosus* has a rich, mushroomy flavor and does not lose its individuality when incorporated into complex dishes. Judging by its appearance on mushroom stamps of ten countries, *L. deliciosus* has a worldwide following. About one hundred countries, the United States not included, have issued stamps depicting local mushrooms. (A small digression: sometime before I wrote this, the U.S. Postal Service decided to keep from circulation a new stamp showing the mushroom cloud of an atomic explosion. I proposed—to little avail—to the authorities that they change the design appropriately and use it to commemorate the more benign interest in garden-variety mushrooms.)

The fact that they "give milk" makes lactarii especially fun because here are mushrooms that *do* something after you pick them—they don't just sit there. I suppose that if I were asked which mushroom genus I would like to have with me on a desert island, my answer would be *Lactarius.* Why lactarii have milk is a puzzle that awaits a solution. Is the milk comparable to the latex of milkweeds or rubber plants, whose liquid is a modified form of sap that helps carry nutrients to different parts of the plant? In the lactarii, the fluid is not found free but is contained within specialized cells. Of course, most mushrooms get along perfectly well without making milk, which certainly clouds the issue. I know for a fact that the milk-like latex can elicit unusual mental associations. An observant Jewish friend of mine had the fleeting thought that lactarii must be dairy-like and, in accord with kosher dietary laws, should not be eaten with meat. It's all right, he soon reassured himself, they are "parveh," neither dairy nor meat!

On a good day in the middle of the collecting season, one is almost certain to run into boletes. This large taxonomic group is distinguished by having tubes under the cap—think of them as a grand alternative to the gilled mushrooms. On the underside of a bolete's cap one sees the openings of the tubes, small pores that give the mushroom a spongy appearance. The tubes are long and narrow and make up a layer of tissue separate from the rest of the cap. The two distinct layers that make up the flesh of the cap and the tubes can readily be seen by splitting the cap.

There are so many boletes that telling them apart is quite challenging. One of the most useful distinguishing features found in many boletes is a kind of "ornamentation" on the stem in the form of netting, dots, or vertical streaks. To show what I mean, I enlist a romantic passage from Tolstoy's *Anna Karenina* that parenthetically, if inauspiciously, illustrates this very point. One of the protagonists, Koznyshev, has just missed the opportunity to propose to Varenka and, instead of speaking of romance, he suddenly turns the conversation to mushrooms:

"What is the difference between a white and a birch mushroom?" Varenka's lips trembled with agitation when she replied: "There is hardly any difference in the cap. It's the stalks that are different." And the moment those words were uttered, both he and she understood that it was all over, that what should have been said would never be said. "The stalk of the birch mushroom," said Koznyshev, who had regained his composure, "reminds me of the stubble on the chin of a dark man who has not shaved in two days."

Varenka was right and, alas, so was Koznyshev, mycologically speaking. The two kinds of mushrooms can be told apart by the "stubble" on the stem of the "birch mushroom" and its absence on the "white mushroom." The birch mushroom is *Leccinum scabrum.* The "white mushrooms" *(Boletus edulis)* are much better known and are called the "king boletes" in the western United States, "porcini" in Italian, "cépe" in French, and "Steinpilze" in German. King boletes have a fine taste and are among the most prized of all wild mushrooms.

Boletes are common all over the temperate zone and are used for food in most of the countries where wild mushrooms are consumed. In North America and Europe, there are hundreds of bolete species, a few of which are found frequently, others rarely. I remember one fall outing of the Boston Mycological Club when two dozen members came back with exactly two dozen different kinds of boletes. Not only was that a numerical coincidence, it must also have been a record for variety encountered in one brief foray.

Boletes vary in size from small, inch-wide caps to dinner plate–sized monsters. Our boletes are dwarfed by much larger specimens, some reaching twenty inches across and weighing several pounds each; these are found in the southern hemisphere, in Australia and the tip of South America. Boletes also display a large repertoire of colors and silhouettes. Many of them are edible, but some are bitter, peppery, or, at best, exceedingly bland. A few are poisonous, but the damage they cause is generally confined to gastrointestinal disorders. Thus, as is the case with all mushrooms, one must get to know the boletes before eating them.

In the northeastern United States one can find a stately-looking bolete similar in appearance to the king bolete and often present in such abundance as to promote fantasies of great feasts. However, this is the bitter bolete *(Tylopilus felleus)*, which earns its name because of its intensely unpleasant and lingering taste. Pity the unwary who does not know the difference and spoils a whole stew by adding this mushroom to the pot. This is a mushroom that promises so much and delivers so very little. As if to make up for the sins of its cousin, one of the most beautiful of all the mushrooms, Frost's bolete, or *Boletus frostii,* also populates the eastern woods. Its cap is of a bright and shiny crimson color that a Japanese lacquer-box artist would dream of. The stem is covered with an intriguing meshwork, often with contrasting areas of yellow, and the pores exude bright yellow droplets. Each specimen appears to be carved with loving care and painted with intense pigments. This species is edible, but it is seldom gathered because field guides recommend staying away from boletes with red pores that turn blue on bruising, as most of the poisonous boletes do. There are exceptions, and *B. frostii* is one of them. Nevertheless, red-pored boletes that bruise blue should be avoided by all but the experienced mushroomers.

Together, the amanitas, russulas, lactarii, and boletes make up a high proportion of the mushrooms in the North American forests. On a typical successful walk, though, you are likely to find many more. Let me introduce a few other kinds that are found frequently. In a forest of conifers, the ground is sometimes covered with clumps of whitish, branched structures that look like marine corals. They are indeed called the "coral mushrooms," or clavarias. They make one wonder how living entities separated by such an immense biological divide could have converged to evolve similar shapes. As do their marine counterparts, coral mushrooms also have a lovely appearance; some species display bright colors such as amethyst, pink, or golden yellow, although most are whitish or tan. They tend to be crunchy (but nothing like the real corals!) and some are used for food, but more for texture than for taste. Some of the coral mushrooms are poisonous, so mushroomers must take the usual precautions.

Mushrooms like moisture, but in moderation, and they are seldom seen where the ground is very wet. Sometimes, however, in a boggy area of the woods, growing out of bright green star moss, one can find small, shiny mushrooms that advertise their presence with colors near neon-light intensity. These are waxy caps *(Hygrocybe coccinea),* so called because they feel waxy to the touch, especially on the gills. Hardly any group of mushrooms can beat the waxy caps in color intensity. Their bright shades of red, orange, pink, and yellow make these the showiest of mushrooms, especially if they are found against a vivid mossy background. One species merits the name of parrot mushroom *(Hygrocybe psittacina)* on account of its bright green colors. These are the flowers of the forest.

On a good day for mushroom hunting, my basket is full to the brim with specimens as I come out of the woods. If I am lucky, I may have found a few good edibles—a large clump of honey mushrooms *(Armillaria mellea),* a few chanterelles *(Cantharellus cibarius),* or some horns-of-plenty *(Craterellus fallax).* Most of my finds are likely to be common to the area and easily identified, some others require detailed work. In any case, I never seem to be disappointed. Even if I bring home nothing for the table, I feel I have been rewarded by the pleasurable sight of mushrooms adorning the forest with a variety of shapes and colors. There is always something—old or new—to hold my attention.

5

A WALK ON THE LAWN

Mushrooms that grow on lawns and pastures occupy a special place in the world of mycology. These mushrooms differ from those found in the forests, except perhaps those at the fringes. This is not surprising, considering how different woodlands and grasslands are with respect to moisture, temperature, wind, and the kinds of plants they support.

Most of the time, mushrooming in open spaces is restricted to just a few days after heavy rains have fallen, whereas since the woods retain moisture longer, forest mushrooms appear over a longer period of time. The most obvious advantage of picking mushrooms in the open is that they are more easily spotted at a distance. I have known mushroom pickers to scour large expanses of fields with their binoculars—quite a different routine from the usual foray into the forest, where one's eyes are drawn to the area just a few yards in front of one's feet. (That is not the only difference: "lawn mycologists" exchange their insect repellent for suntan lotion, and the astute ones know that some suntan lotions also discourage insects.)

Mushrooms tend to grow well on relatively neglected lawns, such as those in playgrounds or cemeteries. A few make it through lush suburban grass as well, as long as the lawn has not been sprayed with fungicides. Golf courses are fine places to gather mushrooms, except for the obvious risk of getting conked by a golf ball. Still, my golfer friends serve a useful function as sentinels for the appearance of mushrooms. Golfers know when mushrooms are on the land because, to their annoyance, they can mistake puffballs for golf balls. Whack! There goes another unwary puffball! News of such calamities comes in handy, because if there are mushrooms in open areas there will probably be a lot more in the forest.

Puffballs that grow in open spaces range in size from the small, less than an inch across, to the elephantine. Specimens with a diameter of forty inches and a scale-bending weight of fifty pounds have been reported in the Midwest and the West. In small towns especially, such finds tend to make the local paper. David Arora, the author of a widely appreciated field guide, *Mushrooms Demystified,* says that "large specimens have been mistaken by passersby for herds of grazing sheep." Of course, "mushroom hunters are more likely to mistake grazing sheep for giant puffballs."

The large puffballs belong to the genus *Calvatia,* which translates into something like "baldy," and *Calvatia gigantea* is the appropriately named champion species. The odds of finding such prize-winning specimens are against you, but basketball-sized ones, ten to twelve inches across, are often found. The mid-sized puffballs, about four inches across, are particularly welcome, because in many people's opinion they have the best flavor of the group.

Small puffballs, smaller than Ping-Pong balls, can also be spotted in open spaces, although they are found frequently in forests. Occasionally, one runs into an old fallen tree covered with hundreds of these small puffballs. Other common puffballs grow directly on the ground, although sometimes you can't be sure, as wood may be buried beneath. Small puffballs, found throughout North America, are among the most common members of another genus, *Lycoperdon.* As you'll remember from Chapter 3, "puffballs" are so called because their spores escape in a puff of air from a hole on the top of the ball. The same ability accounts for the name of the genus *Lycoperdon,* which means "wolf's fart."

Puffballs are edible as long as they are pure white inside and have a firm consistency, somewhat like tofu. Puffballs that are poisonous are usually dark inside, even when young. This rule is not quite foolproof, however: the immature specimens of at least one unwholesome puffball, *Scleroderma citrinum,* are sometimes white inside. It is especially important to realize that individuals may react differently even to mushrooms that are well known to be edible. It seems, however, that not everyone can be restrained when tempted by the bonanza represented by a giant puffball. Puffballs are bland in taste and invite experimentation in the kitchen. A common way of cooking the large ones is to cut them into slices first; then the slices are sautéed, perhaps after they have been dipped in beaten egg and bread crumbs. Some people swear by the French toast technique: dip the slabs of mushroom in beaten egg, fry them lightly in butter, then drench them with syrup.

Many other open-field mushrooms are large enough to be seen at a distance. Some are species of *Agaricus,* cousins of the white button mushroom of the supermarket. Like their domesticated kinfolk, these mushrooms have a ring around the stem and gills that turn progressively darker as dark brown spores are formed. It is easy to tell at a glance if you have a young or old specimen by the color of the gills and by the size of the cap, which expands with age. Many of the *Agaricus* species are good to eat, but some cause gastrointestinal disorders similar to food poisoning. The most common edible type in the Northeast and the Midwest is the *Agaricus campestris,* a species with an emphatically redundant name (it means "field mushroom of the field"). In a small town in the plush dairy region of western Wisconsin, I once saw a magnificent display of edible *Agaricus* nearly covering the lawns of three adjacent lots. These were large, appealing specimens, none of which seemed to be of any interest to the local people. What a waste and what a paradox to find a bounty, untried but readily at hand, in the middle of farm country!

In regions where people have to water their lawns regularly, grass is where you'll find the most dependable and abundant mushroom crops. On the city lawns of San Diego's Mission Bay I have found in profusion the edible "crocodile" mushroom *(Agaricus crocodilinus)*, so named because of its warty cap. Even more common on the coastal plains of California, as well as in the southern United States, is the green-spored parasol *(Chlorophyllum molybdites)*. This species is poisonous and responsible for many of the gastrointestinal mushroom poisonings in North America. The green-spored parasol closely resembles several edible mushrooms, such as the delicious shaggy parasol *(Lepiota rhacodes)*, but, as its common name indicates, its spores have a distinctive dingy-greenish color. This is a unique characteristic—practically no other mushroom in North America has it. One reason people eat green-spored parasols is that they have collected immature specimens, which are often stingy in making spores; in these young mushrooms the gills remain white for a long time, just like those of the edible parasols. The lesson is clear: making a spore print is imperative if you are thinking of eating a parasol; if you cannot determine the color of the spore, the collection should be discarded.

Open spaces are also the habitat of an extraordinary group of mushrooms, the inky caps *(Coprinus)*. Their peculiarity is that their caps self-destruct as they mature. The cap tissue dissolves into a black fluid that falls dropwise to the ground. One of the inky caps *(Coprinus comatus)* is particularly stately looking and seems to lord it over its surroundings. It is called the shaggy mane because of the brown scales that cover its white, elongated, egg-shaped cap. Shaggy manes, which are sometimes found in profusion, are about half cap and half stem in height, and some are taller than twelve inches. Highly regarded for their delicate and delicious mushroomy flavor, shaggy manes must be prepared with dispatch, or else the specimens will liquefy and you will be left with nothing but a black mess. All inky caps should be heated quickly to a high temperature in order to disarm the enzymes responsible for breaking down the tissue.

Shaggy manes put on a mushroom-disappearance act. In preparation for a party, I once picked a splendid fresh specimen, sod and all, and placed it on a plate for my guests to admire. As the evening wore on, the cap of the mushroom did what it was supposed to do: it gradually vanished from sight, dripping black drops from its receding edge as it went. By the time the cap had disappeared, a forlorn-looking stem was all that remained: the perceptive ones among my guests understood it was also time for them to vanish.

Other kinds of inky caps are often found in pastures and lawns, but one species has a distinct preference for the base of an old tree. Year after year, tight clumps of hundreds of specimens will spring up around tree stumps. This species is known as the mica cap (*Coprinus micaceus*) for the glistening bits of tissue that cover the caps of very young specimens. These mushrooms—found in gatherings of hundreds of individuals in various stages of deliquescence—are an unforgettable sight. The season for this species begins very early for New England, March or the beginning of April, and it can be found year-round in warmer climates.

Another common inky cap, the *Coprinus atramentarius,* has an unusual effect that is shared with only a few other species. In some people it causes a highly unpleasant reaction when consumed with alcohol, akin to that produced by "Antabuse," a drug that used to be prescribed to alcoholics to discourage drinking. A pity . . . what good is a mushroom dish without a glass of wine?

Lots of tiny mushrooms, a half an inch to an inch across, found on lawns challenge one's identification skills. They are known as LBMs, or "little brown mushrooms." Some of the lawn mushrooms are hallucinogenic, but they are so tiny that they are seldom collected in sufficient amounts to be effective. However, toddlers foraging on lawns may ingest them, as well as other kinds of mushrooms. I used to be referred calls from the Boston Poison Center, and I would often hear from worried mothers who had found their children clutching mangled remnants of mushrooms and chewing. By the time she called me, however, the mother had usually followed the advice of the Poison Center and, thanks to ipecac, the ingested parts of the mushroom were on their way out.

Not all mushrooms are limited to either woods or fields— some can be found in both. On the wood chips used to mulch flower beds or to cover footpaths in and out of forests, you can find mushrooms sometimes as big as dinner plates. They are likely to be wine caps (*Stropharia rugoso-an-nulata),* so called because of the reddish color of the caps. One of the distinguishing features of this species is a prominent ring on the stem that splits into recurved triangular segments; it looks like an oversized diamond clasp with as many as a dozen prongs. Wine caps have a novel, pleasant taste, vaguely reminiscent of potatoes or turnips. In all likelihood, this species was brought accidentally to North America, probably via wood chips. Wine caps seem to be spreading throughout North America, perhaps from an origin in New England, where they have become increasingly common. This mushroom is cultivated in Europe and may eventually show up in American markets, too.

The *pièces de résistance* of lawn mushrooming are the fairy rings, those mysterious circles or arcs where the grass is greener than its surroundings. In the mushroom season, lots of specimens make an appearance at the outer edges of fairy rings. Seeing a fairy ring from a distance allows you to guess that mushrooms will be found there, a feat that may impress your friends. Fairy rings tend to be regular in pastures and fields, but in the woods they are rarely complete circles, because of irregularities in the terrain. The most common type is called, of course, the fairy ring mushroom (*Marasmius oreades*), also known as the scotch bonnet. It is smallish, no more than two inches across the cap, and is one of the earliest mushrooms to appear in cold climates, emerging almost as soon as the snows melt in March or April. In warm parts of California, it is found year-round and even gives several crops. It is well worth eating.

What makes fairy rings? Elves and fairies used them for dancing, according to folk theories. On the other hand, the people in the Tyrol ascribed the rings to the scorching breath of dragons, while the Irish thought that they arise from milk spilled when the devil was making butter. In general, fairy rings were not considered good things. A Scottish rhyme tells:

> He wha tills the fairies' green
> Nae luck again shall hae;
> For weirdless days and weary nights
> Are his to his deeing day.

Even today, the whole story is not understood—though we now know that fairy rings mark the periphery of an underground mycelium. If you look closely, you will see that the grass grows more densely at the periphery of the ring, probably because the mushrooms make growth-stimulating plant hormones. Inside the ring, however, the grass is somewhat stunted. The puzzling thing is that mushrooms are seldom seen inside the ring. At first glance, this may seem obvious—the center areas have either run out of food or have accumulated substances that inhibit the fruiting of mushrooms from the fungal growth in the soil. But some fairy rings are hundreds of years old (and cover many acres of land, as, for example, in the pastures by England's Stonehenge), which begs the question of why, after so much time, the original state of fecundity has not been reestablished inside the circle.

Circular growth is a predilection of fungi not limited to lawns or to terrestrial fungi in general. Molds on fruits tend to grow in round patches, often with concentric zones; that is how colonies of molds grow on agar-containing Petri dishes. This tendency toward roundness is also seen in the skin lesions of humans and animals affected by ringworm. This superficial disease is caused by fungi growing on bare skin. Here, the "old" growth inside the circle is inhibited by the inflammatory response of the host. Can we make a giant leap, in biology and in logic, and infer that grass responds in like ways to the growth of fairy ring mushrooms? Are fairy rings "ringworm of the lawn," or is ringworm "fairy ring of the skin?"

Whether you are foraging in forests or lolling about on the lawn, you will find mushrooms only when the time is right. Prolonged rains and mild temperatures are the optimal weather conditions for mushrooms in the temperate zone. The most productive regions are those where precipitation is abundant and reliable, such as the Pacific Northwest. That is in fact the North American mushroomers' paradise, where harvests are often measured by the bushel. The Rocky Mountains also yield abundant harvests, as long as the temperature is not too low. The East, much of the Midwest, and parts of the South have a great variety, but mushrooms are generally less abundant in these regions and people usually have to work harder for their mushrooms. In the Northeast, optimal conditions occur every five years or so. Most other years, mushroom pickers pray for rain. When the weather in late summer and early fall is good, it's very good, and when it's bad, it's terrific!

The mushroom season peaks at different times in various parts of North America. In the North, from the Atlantic to the Rockies, there is a brief flush of mushrooms in the spring, but the height of the season is the late summer and early fall. The season is similar in the mountains of the West, but it is longer in the South, where mushroom hunters may be kept busy roughly from February to November.

Certain mushrooms can be found over a wider period of time than others, but many kinds are either early or late bloomers. Some species—for example, the oyster mushroom (*Pleurotus ostreatus*)—tolerate surprisingly low temperatures. Three or four times a winter, whenever there was a thaw, I used to harvest a nice collection of oyster mushrooms from the stump of a mulberry tree in my back yard in Massachusetts. The mushrooms must have finally run out of food as the stump shrank in size. Picking up this kind of tip is a good reason for seeking out fellow mushroom lovers. You can learn a lot about the seasonal occurrence of individual mushrooms by joining an amateur club; because these clubs can devote a great deal of manpower to a study of the local mushroom habitats, they are often a better source of information about your area than a professional mycologist or a guidebook that covers your state or geographic region. Mushroomers like to keep lists, although not as formally as bird watchers do; there is no organization that promotes keeping a mushroom "life list," nor are alerts given for the sighting of a rare mushroom species. For many people, however, getting together with like-minded enthusiasts is not just a means of sharing information—it is one of the joys of having a hobby. As the next chapter attests, the science of mycology may be associated with university biology departments, but the art of mushrooming is the province of amateur clubs.

6

MUSHROOMERS UNITED

As time goes by, mushroom hunting is becoming more and more popular in many parts of the United States and Canada. The number of people involved is hard to estimate, especially because mushroomers range widely in the depth of their interest, from those who occasionally venture eating a wild species to those who want to learn every species in sight. One indicator of burgeoning interest is the impressive increase in the number of mushroom clubs in the 1970s and 1980s. There are now very nearly a hundred clubs in the United States, about equally divided between east and west of the Mississippi, and eight in Canada. Washington and Oregon, with eighteen clubs between them, are the most mycophilic states *per capita*. Elsewhere in North America, although mushrooming is still regarded as a bold and unusual avocation, its practitioners are no longer placed in the category reserved for amateur parachutists or daredevil bungee jumpers. The day may soon arrive when it will be as easy to find a mushroomer in certain parts of North America as it is now in France, Italy, or Germany.

Nowadays, many people "get into" mushrooming with the same ease as they sign up for bird watching or wildflower identification. Most become involved by joining a club or going along on a mushroom hunt with a knowledgeable friend. Before this onrush of popularity, when mushroomers were few and far between, there usually was more to the story of one's mycological initiation. A good example is the way Margaret Lewis, the late *grande dame* of Boston mycology, got started, but before telling her story, I should introduce you to her.

For many years, Margaret Lewis taught a course on wild mushroom identification in Cambridge, Massachusetts, and many New England mushroomers owe their start in mycology to her. Margaret weighed all of 92 pounds, but heavy was the impact she had on her audience. She put on quite a spectacle, flitting excitedly from one subject to another in seemingly random fashion. Pausing in midthought, she would turn to the class and ask a challenging question, then suddenly veer away from what she was talking about as if overcome by the beauty of the next idea. She held her audience spellbound with unexpected and sometimes implausible-sounding stories delivered with utmost assurance and vivid gestures. More than one of her students signed up for her course again the following year. For her memorial in the *Boston Globe* (May 27, 1987) I wrote: "If I were casting for a play, I would have chosen her for the role of the duchess. Like a duchess, she was imperious and self-assured, and although she liked most people, you had to earn her attention."

Here is what Margaret told me about how she got started with mushrooms. In the early 1930s she used to make sketches of wildflowers whenever she could. She was also an avid hiker and would outpace her husband (called "Santa," for reasons unknown to me) in arduous climbs of New Hampshire's White Mountains. While waiting for him to catch up, she would take out her sketchbook and start drawing. On one occasion, she found herself surrounded by mushrooms and decided to sketch them. Not knowing what she had drawn, she took the sketches to the Boston Mycological Club, where she got a frigid reception. In the early 1930s, most members of the club were elderly people who preferred to conduct their business in a formal atmosphere (ladies used to go on forays wearing gloves!). Undaunted by the initial cold shoulder, Margaret eventually joined the club and proceeded in time to become one of the leading lights in American amateur mycology. Her contributions included detailed experiments with various ways of preserving different species and inventing new dishes that brought out the special qualities of wild mushrooms.

My first exposure to mushrooming came about by good luck. For mushroom hunters, 1968 was a memorable year in the Northeast: it rained a lot and mushrooms seemed to pop up everywhere. One day my wife called me at work to say that she had gone out to the back yard and collected a paper bag full of specimens. (This profusion of mushrooms was surely propitious, because normally I consider myself lucky if I find more than a couple of specimens in the yard.) Friends of ours living in Oregon had described the pleasures of wild mushroom picking, she reminded me, so on the way home would I please stop at a bookstore and buy a mushroom book? We could then find out what she had picked. I selected Alexander Smith's *Mushroom Hunter's Field Guide* and gave it to her to proceed with the identification. Not five minutes had passed when, sensible person that she was, she realized that this was going to be difficult and gave up.

I responded more positively, perhaps because I had lived for four decades without a hobby and apparently had been waiting for one to come along. In my early days as a microbiologist I had worked with microscopic fungi—molds and yeasts—and though this did not seem to have much to do with identifying wild mushrooms, at least I had visited the realm. As I leafed through Smith's book, I became engrossed in the pictures and marveled at the great varieties of unfamiliar species. I recall finding the word *bolete,* which struck me as funny because *boleto* is Spanish for "ticket." I also looked up *Amanita* in the book, a name I had previously learned, probably by osmosis. I soon found out, however, that most of the mushrooms in the book looked, well, like mushrooms, and that telling them apart would be hard work indeed.

The next weekend I saw an announcement in the *Boston Globe* that the Mycological Club was sponsoring a walk in nearby Milton, Massachusetts. Amazingly, that was the only time such an announcement has appeared; at least in the 1960s, the club was leery of "recreational" mushrooming and was not bent on proselytizing. So I went. I found a couple of dozen ladies and gentlemen armed with baskets, ready to be sent into the woods on command. Introducing myself as a newcomer, I was received a bit more warmly than Margaret Lewis had been in her day.

The garb of the forayers was not as formal as in the past, although I was soon greeted by an elderly gentleman in shirt, tie, and jacket . . . and sneakers. Recognizing me as a newcomer, he told me to tag along with him. Soon he asked me what mushrooms I was particularly interested in. I think I was able to mumble something about "the amanitas," biting my tongue to hide my ignorance, but I had enough wits about me to ask him what *his* interests were. "Thoreau and mushrooms," he said loftily, as if it should have been perfectly obvious. Later I found out that Thoreau did indeed have something to say about mushrooms, not all of it positive. (In one entry in his *Journals*, he wrote: "Pray, what was Nature thinking of when she made this? She almost put herself on a level with those who draw in privies.") It didn't take me long to discover that my mentor was only one of a corps of real characters, each with interesting quirks. Perhaps half of the club members were Yankees, a breed that most people only occasionally encounter in Boston, expectations to the contrary. I asked, with proper humility, if I could join the club. I was welcomed and began to attend

On the Trail of a Good Book

While I was visiting my godson Robert in Norwich, England, he offered to take me on the rounds of antiquarian bookstores in search of mushroom books. We soon learned that we should try our luck in the nearby small town of Bungay. Indeed, there we found a bookstore that had some finely illustrated mushroom books from the eighteenth and early nineteenth centuries. Most were beyond my financial reach, but one made me hesitate. It was not a printed book but rather a bound collection of fine watercolor sketches. It caught my eye because of a possible American connection. The volume did not carry the author's name, but inscribed on an inside page were the name of a Philadelphia lady and the year 1825. In the beginning of the last century, mycology was not a major occupation in the United States, so anything that might be relevant to early American mushrooming was of potential interest.

I decided to buy this book and learn its place in American mycology. About all I knew about this period was that its most important figure was a priest of the Moravian Church, Lewis von Schweinitz, who had studied fungi both in Europe and in North America. The cover of the book carried the imprint *Icones Mycologicae Niskiensis* (Pictures of mushrooms of Niesky). Niesky, it turns out, is a town in eastern Germany, near

continued

the Polish border. It was founded by the Moravian Brothers, the order to which von Schweinitz belonged. Was I on the trail of a link between the book and von Schweinitz?

I knew where to go for help—namely, to an expert in botanical bibliography, the Harvard mycologist Donald Pfister. Don took it upon himself to research this book, and in short order he called to tell me that he had indeed found a strong connection to von Schweinitz. Don is not one to get readily carried away, but I could hear excitement in his voice. He had compared the handwriting of von Schweinitz (from manuscripts at Harvard's Farlow Cryptogamic Library) with that of the legends in the sketch book, and they matched! In addition, in some of the legends, the Latin name was followed by the word *nobis,* denoting a new name proposed by the author. Indeed, all these species names were coined by von Schweinitz.

So, by accident, I have been given the gift of participating in the rescue of an important mycological work. It was known that von Schweinitz had made four volumes of sketches, but only two of these had been previously identified. My find reduces the missing number to one, which awaits discovery by some other fortunate collector. Meanwhile, "my" Schweinitz is safely ensconced at Harvard.

both the Sunday walks and the identification meetings held on Mondays. I was somewhat taken aback, however, when after having attended assiduously for over three years, I was told by one elderly gentleman after a meeting: "Nice to see you, come again!"

From my time spent at the club forays and identification meetings, I learned that several of the members of the club's identification committee could match Margaret Lewis's vast knowledge of mushrooms, or come close to it. One of these people, Ruth Lever, had a legendary memory for shapes and forms and would remember mushrooms she had seen but once, perhaps twenty years before. There seemed to be no mushroom that could escape her identification skills. Most of the active club members at the time I joined were elderly people who had the time to dedicate themselves to this pursuit. Currently, many younger and otherwise busier people are becoming involved in mushrooming. This, however, does not mean that the present club members are any less picturesque or less committed to the challenges of identification.

The principal activity of mushroom clubs everywhere is to sponsor walks during the collecting season. Members gather at promising sites, are given a few instructions regarding likely finds and how not to get lost, and are let loose to traipse through the woods. Everyone returns at a prearranged time and gathers around a display table, where the specimens are laid out. The identified species are laid on tables with slips of paper giving their names. Animated discussions regarding identification regularly ensue, with everyone more or less deferring to the gurus in the group.

There is a special etiquette to collecting in groups. For example, the privacy of one's collecting basket is sacrosanct, and no one except an artless beginner would dare to pick a specimen from somebody else's basket without permission. On the other hand, once specimens are taken out of the basket or laid out on a display table at the end of the foray, all manner of smelling, tasting, touching, and squeezing is allowed. Sharing a particularly rich harvest of prized edibles with others is strictly up to the finder, but holding on to a large take for one's own use is not only tolerated but well understood. Always readily shared are recipes, anecdotes, warnings regarding poisonous species, and general information about appropriate times and sites for picking.

Mushroom picking is accompanied by considerable ritual, although the details vary from culture to culture and from person to person. One of the most ornate of rituals accompanied the hunting of matsutake in Japan—at least in the days when this species was still abundant there. Among the nobility, considerable care went into the dress and coiffure prescribed for these expeditions. The men wore tight-fitting green leggings *(patchi),* which were freely displayed when they would secure their robe about the waist to leave their legs free. The women wore embroidered gaiters of white or purple silk, said to "flutter like anemones" when the kimono was kilted through the obi sash for freer movements. On returning to the picnic site, those who had been particularly successful were much congratulated, while those who made a poor harvest apologized with mock humility for their stupidity and unworthiness to accompany such a distinguished party. The matsutake were then toasted over a pine fire and enjoyed with soy sauce and vinegar.

Comparable ceremony was associated with mushrooming in parts of Europe as well. For example, in the great Polish epic, *Pan Tadeusz,* by Adam Mickiewicz, there is a detailed description of a mushroom hunt by an aristocratic party. The "solemn rite of mushrooming" required that some members of the company put linen dresses over their coats. Some wore "great hats of straw upon their head, like pallid souls in the purgatory of the dead." There is a gleeful side to this imagery, for in the words of the host,

> Who brings the finest mushroom to the board,
> The loveliest girl to sit with I'll award,
> His choice; but if a lady gather it,
> The handsomest young man shall next her sit.

In the absence of established rituals, people seem to develop some of their own. I can vouch for the fact that most habitual mushroomers, myself included, pay an inordinate amount of attention to the garb they wear, no matter how informal, as well as to the basket and other paraphernalia they take into the woods. Of course, concentrating on one's gear also fulfills functional requirements—there is always the possibility of rain or insect bite—but I think it goes beyond mere practicality. Is one basket "luckier" than another? Do certain clothes ensure good picking?

Once in the woods, everyone seems to follow an established although highly personal pattern. Some people wander far afield and take in the grand view, others stick to a small area and carefully examine every inch of it. Personal interactions vary all over the map: there are both solitary and gregarious forayers. For me, mushrooming is and isn't a solitary activity. I often enjoy being alone in the forest, but at other times I like the company of people with the same interest. There are occasions when I feel more comfortable being in the woods with others, or maybe I just like sharing the joys of the hunt. None other than Sigmund Freud, who was an avid mushroomer, developed an elaborate ritual for hunting mushrooms with his children and grandchildren. Freud's son, Martin, wrote in *Sigmund Freud: Man and Father* (New York, 1958):

> We had no fear. Father had taught us much about fungi, and I do not recall an occasion when we brought a poisonous species for him to inspect and pass as safe . . . Our attack on the mushroom was never haphazard. Father would have done some scouting earlier to find a fruitful area; and I think one of the pointers he used was the presence of a gaily coloured toadstool, red with white dots, which always appeared with our favourite, the less easily seen *Steinpilz* . . . Once the area had been found, father was ready to lead his small band of

troops, each young soldier taking up a position and beginning the skirmish at proper intervals, like a well-trained infantry platoon attacking through a forest. We played that we were chasing some flighty and elusive game; and there was always a competition to decide on the best hunter. Father always won.

Freud's hat was an important part of the collecting ritual:

. . . usually a grey-green velour hat with a wide dark-green silk ribbon. One sees these hats occasionally in England, where they are called Austrian hats. When father had spotted a really perfect fungi specimen, he would run to it and fling his hat over it before giving a shrill signal on the flat silver whistle he carried in his waistcoat pocket to summon his platoon. We would all rush towards the sound of the whistle, and only when the concentration was complete would father remove the hat and allow us to inspect and admire the spoils.

And further:

When they went collecting mushrooms he always told them to go into the woods quietly and he still does this; there must be no chattering and they must roll up the bags they have brought under their arms, so that the mushrooms shall not notice; when father found one he would cover it quickly with his hat, as though it were a butterfly. The little children—and now his grandchildren—used to believe what he said, while the bigger ones smiled at his credulity; even Anna did this, when he told her to put fresh flowers every day at the shrine of the Virgin which was near the woods, so it might help them in their search. The children were paid in pennies for the mushrooms they found, while the best mushroom of all got a florin. It was the quality, not the quantity of the mushrooms that mattered. (From a journal entry of December 1921 cited in *The Sigmund Freud and Lou Andreas-Salomé Letters,* edited by E. Pfeiffer et al. [New York, 1972])

Surely some of the fascination of eating wild mushrooms is that it is like playing with fire. There is a measure of danger, one that even the most experienced mycologists acknowledge, if not always aloud. I believe that my own fears are usually under control, and I don't fret visibly when eating a reputable species that I have eaten before. My "warning signal" flashes when I am confronted with a species new to my palate, however. I imagine all that could happen, but more often than not I end up eating a small amount. Perhaps it is the thrill of the unknown, similar to the "culinary adventure" I had in Fukuoka, Japan, when I was encouraged by a Japanese friend to eat fugu, the highly poisonous blow-fish. I knew that this fish contains a powerful neurotoxin in its viscera, the least trace of which is lethal and must be removed with great care. Only licensed fugu cooks are en-trusted with this dangerous procedure. As I was timorously sampling the fish, my host reassured me by quoting an old saying: "Those who eat fugu are stupid but those who don't eat fugu are even more stupid." So with wild mushrooms.

People often need to joke about the dangers they face.

Little wonder that several mushroom clubs call their yearly celebration "The Survivors' Banquet," or that certain walks are called "The Optimists' Foray." Ditties and mild jokes are often exchanged, especially between experienced mushroomers and beginners. Most of the humor may not seem particularly funny out of context, but it does reveal something of how people confront danger. Consider the novice's question: "Can you eat this kind?" Answers the experienced mushroomer: "Yes, of course. You can eat any mushroom once." Or the ditty: "There are old mushroom hunters and bold mushroom hunters, but there are no old, bold mushroom hunters."

Most people who get started in mushroom identification do not stay the course and lose interest when they realize how difficult an undertaking it is. Beginners often express a sense of frustration at the confusing variety of forms, the bewildering Latin names, and the seemingly perverse tendency of taxonomists to change the names, sometimes in contradictory fashion. Among those who continue in this pursuit, some have an uncommonly acute visual memory and are able to distinguish a large number of mushrooms with ease. A few rare people are so gifted that they become highly knowledgeable within a couple of seasons. Others add their own special flourish to this endeavor. I am thinking in particular of Milton Landowne, a retired medical researcher who took up mushrooming later in life. His passion for mushrooms has been so contagious and his powers of persuasion so extraordinary that on at least two occasions when he had gotten lost in the woods, he talked the strangers who rescued him into joining the mushroom club!

As well as arranging walks in the woods, most clubs hold technical and gastronomic meetings and banquets and publish newsletters containing recipes and other bits of "mycellanea." Two major national forays are held yearly in different parts of North America (see the list of "Resources" at the end of the book). Each lasts three days and attracts some three-hundred people from all over the United States and Canada. A number of well-attended regional forays are held as well. Professional taxonomists also attend these events, partly because they are willing to teach amateurs but also because they want to examine the large number of mushroom specimens collected. Every year, specimens are found that require considerable research. Are they new to science, or are they rare species that have been described before?

In addition to the professional mycologists, a number of highly gifted amateurs have amassed such knowledge that they can readily be counted among the experts. Luckily, many of these people have a strong interest in helping beginners and have played an active role in the success of local mushroom clubs. The New York club comes to mind, having been founded by several highly dedicated mushroomers, the composer John Cage included. A current leading light of this club, Gary Lincoff, has written several books and field guides on mushrooms and conducted mushroom expeditions all over the world.

Banding together in an organized way for the purpose of

mushrooming is a fairly recent phenomenon. It all began in England in the second half of the last century. The first organized mushroom hunts were held by the Woolhope Naturalists' Field Club, founded in 1851 and named after a small town in the west of England, near Hereford. The membership included many clergymen and physicians. Its guiding spirit was a Dr. Henry Bull, who, having been interested in the higher fungi, invited the club to its first mushroom foray in 1868. It appears that the now widely used and somewhat bellicose term *foray* was introduced on this occasion. Twenty-one people attended the ground-breaking occasion.

The Woolhope fungus foray became a yearly event that, in time, was attended by as many as sixty people. Excursions took place during the collecting season over several weekdays. Saturdays were not convenient for most members, and the clergymen were otherwise occupied on Sundays. Most of the members lived nearby, but eventually mycological notables from farther afield, including the Continent, joined the excursions. Members, in black suits and top hats, were picked up from fixed rendezvous points and transported to the hunt in horse-drawn wagons. The distance traversed during the outings was usually about eight miles. Often the local squire would play host at a meal at the local hotel, the Green Dragon.

This unique event, apparently the only organized mycological excursion anywhere at the time, served as a model for other clubs in Britain and on the Continent and, eventually, in North America. The activities of the Woolhope Club are vividly recorded in the writings of Worthington Smith. His accounts were published in a national magazine, the *Gardener's Chronicle*, which made the Woolhope Club renowned all over England. Here is an example:

> Dr. Bull dispensed the savoury and steaming viands with his own hands to the fifty-two diners, from bowls of rich fungus soup. It really was a treat for all who understood physiognomy (as do all the Woolhopians), to see the unmistakable external marks of internal gastronomic satisfaction suffuse the delighted faces of the recipient as they each consumed the precious and Elysian fungoid comestible dispensed to them from the safe hands of the Doctor.

Mordecai Cooke, a leading British mycologist of the second half of the nineteenth century, attended many of the meetings of the Woolhope Club, as did most other British mycologists. Smith writes:

> Dr. Cooke, furnished with a large leathery travelling trunk (in place of a hand basket or tin collecting case) was one of the first to arrive in the Forest. By 4 o'clock the Doctor's phenomenal portmanteau was full of funguses. Where one generally looks for a tooth-brush might be found a Phallus, in place of a sponge was a bloated Boletus, in lieu of writing paper sheets of dry-rot. Shirts were shirked, and fungi

both fresh and frouzy were in all the compartments of the valise. No one but an advanced fungologist could so treat a portmanteau.

Smith communicated in great detail the activities of the Woolhope Club, including the departure of the forayers:

And so, with many thanks to the kind host and hostess, the party drove off, in the black darkness of the evening, to the Titley station. Owing to the nature of the vehicles, and the quality of the quadrupeds, the party was late at the station, but such is the respect with which fungus-men are held in Herefordshire, that the station authorities detained the train for six minutes, till the arrival of the party.

With the advent of Saturday, the visitors were all dispersed, and the ancient city of Hereford resumed its accustomed serenity. Doctors again breathed freely, all fears of fungus poisoning vanished, the eccentric individuals in thick boots and gaiters, carrying suspicious baskets and candle-boxes, had departed; the locusts had spared the city, and in a few hours Sunday would arrive, and with it the opportunity for thanksgiving for providential deliverance.

In the late nineteenth century, with the great wave of interest in natural history, mushroom clubs were founded in various cities of the United States. Most of these clubs disbanded in the first half of the twentieth century. Of these early clubs, the Boston Mycological Club is the oldest one remaining. It was founded in 1895 and it's worth noting that within two years of its inception it had 430 members, which is more than it has ever had since. Dues were one dollar, but membership was not automatic and had to be approved by a committee. The club held weekly forays during the season, and members usually traveled to the site on trolleys. As I mentioned already, members were nattily dressed, both for the forays and for the weekly identification sessions held on Mondays at the impressive Horticultural Hall in Boston.

Much has changed since then, but some early habits have survived among mushroom hunters. It's likely that Worthington Smith or Mordecai Cooke would feel at home in a present-day foray of one of the many mushroom clubs in North America. Mushroomers include people of all ages and occupations, but they are still conspicuously idiosyncratic. The Boston Mycological Club's founder and first president, Julius A. Palmer, wrote in 1889 in the *Boston Journal*:

> There is a large, white mushroom growing from the elms on Boston Common, the gathering of which brought me at least notoriety: I dislodged it with a long pole with a knife at the end and by practice could catch the cluster on its fall in my hat. Such proceedings secured an attentive if not always a respectful audience at almost any hour.

In the early years of mushrooming as an organized activity, amateurs from all over the United States contributed a great deal to the knowledge of higher fungi. The neat and careful documentation of some of the participants is evident from the 1902 field notes of a Boston mycologist, George Fessenden (Figure 6.1).

Amateurs (that is, people not formally trained in mycology or deriving a livelihood from its study) and professional mycologists communicated and exchanged specimens assiduously. As a result, many species were named after the amateurs who had found them. In those days the distinction between amateurs and professionals was blurred. In France, for example, pharmacists were trained to distinguish edible from poisonous species for the public, and several eventually became famous taxonomists.

The formidable manpower contributed by amateurs has been put to good use. An example are the mushroom censuses that have been taken over the years. From this work it was concluded that in northern European countries, such as the Netherlands, the number of mushroom species appears to have declined in the second half of this century, possibly because of the increase in land devoted to farming. Records of the occurrence of mushroom species kept by amateur organizations in the United States could be used to study similar habitat changes here. The Boston club has a particularly accurate and extensive set of records. They were compiled by the late Ruth Lever, who recorded the species found at club forays with loving care. Over a thirty-year period, Ruth missed practically none of the club's outings, including those held in pelting rain. Her notes allowed us to question whether there had been a decline in the number

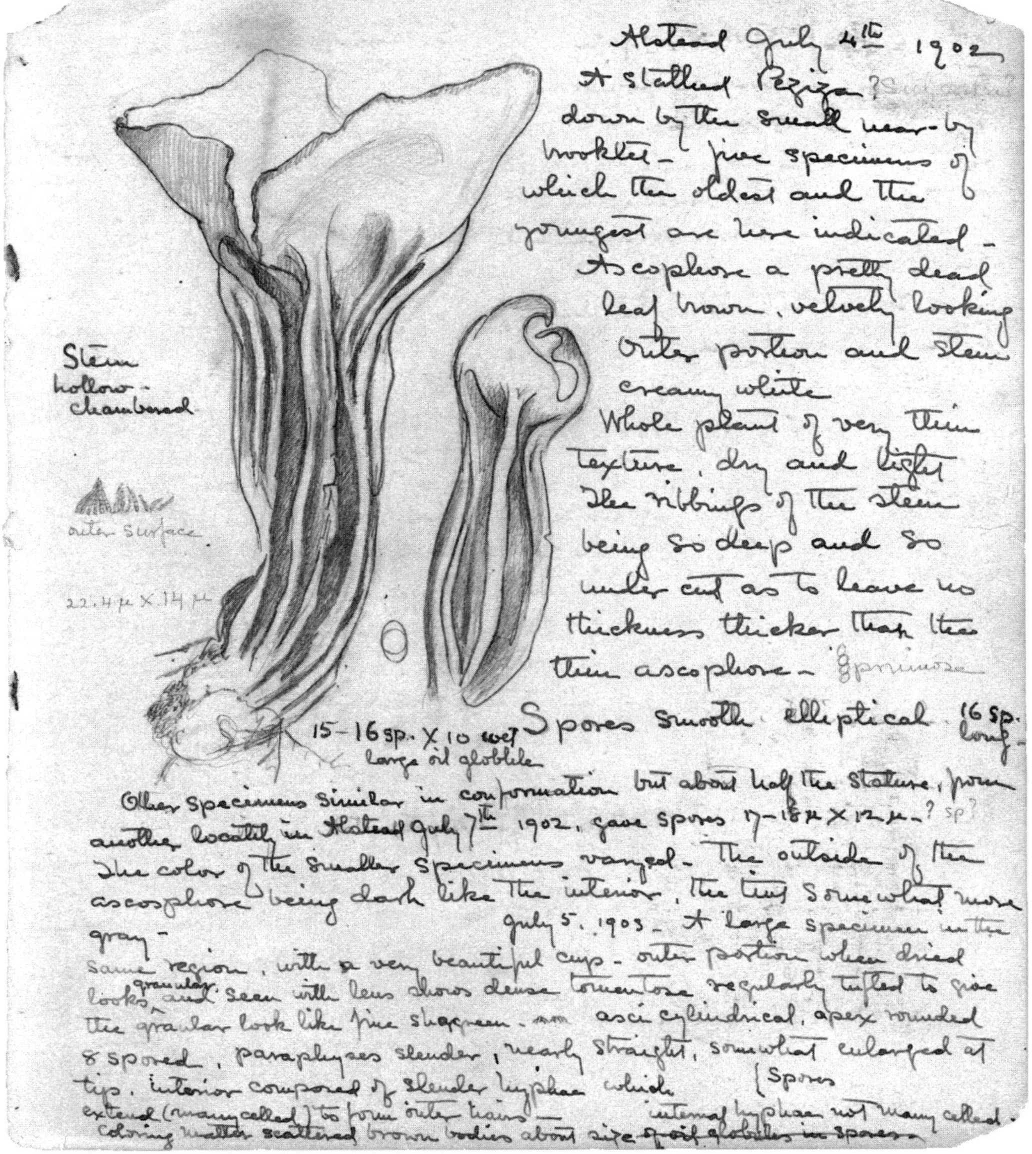

6.1
George Fessenden's field notes, 1902.

of species reported over an eighteen-year period. The answer was no: the incidence of some 500 species had not changed perceptibly between 1966 and 1983. Most of the mushrooms on the list are found in the woods, and thus they serve as sentinels for damage to forests. Does this mean that the relevant habitats of New England have not undergone the same massive alterations seen in northern Europe? Perhaps this is true, thus far. More data need to be gathered over longer periods of time before we can arrive at any definite conclusions. Amateur mushroom hunters, given their number and their enthusiasm, are in the best position to contribute to such studies.

Not the least of the possible reasons for "getting into mushrooms" is the opportunity to "get into nature." If you venture out into the woods or across the fields to search for mushrooms, you will learn something about biological diversity, and perhaps you will learn enough to sound an alarm when this diversity is threatened. Not every mushroom hunter is mindful of it, but every mushroomer is a field biologist, untrained though he or she may be in the intricacies of the science. There is no point in being too modest about it, even if the knowledge acquired is not couched in professional language or is destined to remain within one's mind. Most mushroomers I know do not readily talk about this aspect of their avocation; they are more comfortable discussing the ways that mushrooms satisfy the palate or the eye. I am not fooled by this reticence. I'm sure that my fellow gatherers are urged on by a wish to comprehend their natural surroundings.

The experience of nature to be gained through the hunt is an intimate one. There is a difference between strolling through the woods and being there for a purpose that expressly relates us to the natural domain. The very act of going into the forest to look for mushrooms affirms a connection with the living world that is not often put into words. I feel grateful to the mushrooms for allowing me to make this connection.

Takegari ya yoku kara michi ni fumimayoi

Mushroom gathering;
From greediness
We lost our way.

Jiraku

CULINARY TALES

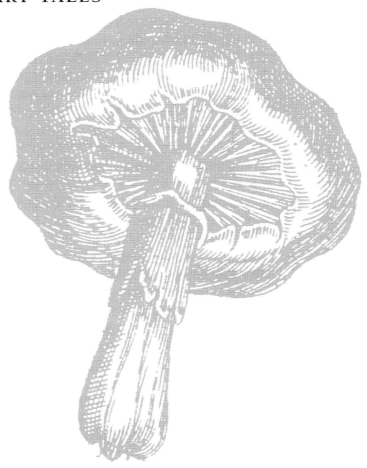

7

A MATTER OF TASTE

Mushrooms, like blueberries, raspberries, and most other foods, taste better if you pick them yourself. Questions of freshness aside, I suppose this preference for "pick-it-yourself" food is a bias left over from our foraging instinct, an impulse so strong that it may cloud our gastronomic impartiality. Fellow mushroomers have proudly presented me with dishes made from specimens they had collected that, to my mind, tasted like buttered balsa wood or sautéed old tennis shoes. Yet I would swear that my own harvests always produce culinary delights.

How can we be objective in such matters? What should be our standard? My wife used to say that it is worth the trouble of picking and cleaning wild mushrooms only if they are at least as flavorful as the white button mushroom found in stores (*Agaricus bisporus*). The trouble is that the *bisporus* flavor, while widespread among mushrooms, is only one of many mushroom tastes, and this makes comparisons difficult. Mushrooms may have the pungency of truffles, the spicy aroma of matsutake, and the seafood-like taste of oyster mushrooms. In texture, they can be crunchy, glutinous, chewy, or slightly stringy (like white chicken meat).

The majority of wild mushrooms flunk any gastronomic test because, although they may be nonpoisonous, they simply do not have much flavor. At best, they add an interesting consistency to a dish, but otherwise they taste of the butter, onions, and garlic in which they are cooked. On the other end of the spectrum are truffles, morels, boletes, chanterelles, and black trumpets, all of which easily outshine the store-bought white buttons. These are the true delicacies that give wild mushrooms their good name. But what about other edible mushrooms, such as the chicken mushroom, the oyster mushroom, or the increasingly available shiitake? Where should they be ranked?

Food researchers and marketers are naturally interested in determining what foods will appeal to consumers, and they have developed quasi-objective ways of "measuring" flavor. How strong, or distinctive, is one flavor as compared with another? "Tasters" are asked to sample different foods and record their reactions either after sniffing or tasting them (in some cases with their noses blocked by nose-clips). Swallowing is not encouraged. The threshold of detection is then determined by having the tasters sample flavors that are more and more dilute. In a 1978 Finnish study, Rakel Kurkela and Eila Matikainen reported the results of testing eight kinds of mushrooms with seven volunteers. One of these people turned out to be consistently less sensitive than the others, but the rest were quite uniform in their response. The lower limit of detection was achieved with samples diluted about a thousand-fold, with a roughly five-fold variation between the strongest and the blandest species. The mushrooms compared favorably in aromatic strength with smoked fish and spices such as marjoram, and they handily outdid tomato concentrates and instant coffee.

Mushrooms are usually cooked with butter or oil or are incorporated into complex recipes, and they rarely stand alone as the main nutritional contribution of a meal. For the record, most mushrooms are 85–95 percent water by weight and have a protein content similar to that of spinach or potatoes. They are a pretty good source of minerals as well as niacin, riboflavin, and other B vitamins. Fresh mushrooms provide about twenty calories per cup with nearly no fats or cholesterol. Most wild and cultivated mushrooms readily absorb the fat they are cooked in, though, so if they are to be used as a base for low-calorie dishes they should not be sautéed but simmered in broths or cooked in soups. No matter, one cannot be unhappy when eating mushrooms, to paraphrase an old Italian saying about eating pasta.

Which species are to be included in the Pantheon of Great Edibles is a matter of taste and geography. Europeans and North Americans seem to agree on which are the most desirable species. When it comes to gastronomy, people from both sides of the Atlantic speak a similar mycological lan-

Edible Mushrooms of North America and Europe

A selection of twelve great edibles, in rough order of gastronomic repute:

- King bolete *(Boletus edulis)*
- Chanterelle *(Cantharellus cibarius)*
- Morels *(Morchella esculenta* and others)
- Horn of plenty, or black trumpets *(Craterellus cornucopioides, C. fallax)*
- American matsutake *(Tricholoma magnivelare)*
- Parasols *(Lepiota procera, L. americana, L. rhacodes)*
- Meadow mushrooms *(Agaricus campestris* and others)
- Chicken mushroom *(Laetiporus sulphureus)*
- Hen-of-the-woods *(Grifola frondosa)*
- Oyster mushroom *(Pleurotus ostreatus)*
- Honey mushroom *(Armillaria mellea)*
- Blewit *(Clitocybe nuda)*

guage. The list of the great edibles is usually headed by cep or porcini *(Boletus edulis)*. Truffles should be listed separately, because the most delectable varieties, such as the black truffle of Perigord, are not native to North America, although some desirable species (e.g., the Oregon white truffle) are found here. There is also a great variety of edible mushrooms in South America, Asia, and Africa, but many of the species that are considered choice in those parts of the

In rough order of gastronomic repute:

Eight Great Edible Mushrooms of Japan

- Matsutake *(Tricholoma matsutake)*
- Shiitake *(Lentinus edodes)*
- Maetake *(Grifola frondosa)*
- Enoki *(Flammulina velutipes)*
- Shimeji *(Tricholoma aggregatum)*
- Hiratake *(Pleurotus ostreatus)*
- Kikurage *(Auricularia fuscosuccinea)*
- Nameko *(Pholiota glutinosa)*

(All but matsutake are cultivated and readily available in food stores.)

world are not found in North America or Europe and vice versa.

As with all such matters, mushroom tastes differ among cultures, and certain species are highly regarded in some parts of the world and avoided in others. The Russians eat a number of mushrooms which in the United States are considered insipid or even poisonous. An example is *Paxillus involutus* (which does not have a widely accepted common name), a species that has been responsible for serious intoxications, including several fatalities, but which is eaten boiled or pickled in Russia. The method of preparation may have something to do with removing toxic compounds. Likewise, the *Russula rubra,* which is sold in the markets of northeastern China, has such a fiery peppery taste that it would be unacceptable to most people in the West. In the south of Chile, a coral mushroom locally known as "mouse foot" *(pata de ratón)* is highly esteemed, whereas elsewhere it is considered to be nearly tasteless. Such distinctions may also reflect variations in the flavor of the same species growing in different regions.

Collecting for eating is different from collecting for study. Obviously, the contrast begins with one's attitude: if the focus is on edibles only, all other mushrooms encountered along the way are just "specimens." Gathering now becomes harvesting, and, to save time and effort, collecting techniques need to be adjusted accordingly. Instead of picking the mushrooms whole, even making sure to preserve bits of dirt and wood attached to the base for the sake of identification, the harvester cleans away all detritus before placing the specimens in the basket. If different edible kinds are collected, they should be kept apart from each other in the basket. Many of the wild edibles are delicate and should not be piled one on top of the other. Place them gently in brown paper or waxed paper bags. Plastic bags are to be avoided, as they trap moisture in the bag, which quickens spoilage.

Where and when to pick edibles varies with the region, so you must learn their geographic and ecological preferences. Some mushrooms are found only in specific habitats, others have broad distribution. On one extreme, morels are cosmopolitan and occupy a wide range of habitats. They can be found in apple orchards, deciduous woods, dying elms, burnt areas, or even suburban lawns. The one constant factor is that morels are found whenever spring-like conditions come to a region: the warmer the climate, the earlier. Morel hunting is a hit-or-miss affair. You can go for a long time without finding a single one, at long last to be rewarded by a collection of fifty or more specimens in a small area. I had an exciting moment some years ago when my neighbor found three beautiful specimens growing on the grass in the middle of his driveway and came over to ask me what those strange things were. The next year he had the whole driveway paved over, a frightening thing to do and hardly conducive to good neighborly relations.

The upper Midwest is morel country, and the morel is the state mushroom of Minnesota. There, and in nearby states such as Iowa, Illinois, and Wisconsin, morel collecting is done with an ardor otherwise rarely found in North America. Local papers and radio stations report on the status of morel picking, and sporting contests are held for the largest number collected in a given time span. The winning number of specimens collected in a two-hour period usually surpasses two hundred. I am not sure where this "morelmania" comes from, but it belies the usual notion that Americans are "mycophobic" through and through.

People who went morelling as children and grew up to

become mushroomers have happy memories as well as experience to fall back on. The experience of a friend, Ruth Halliwell, a native Oregonian, is typical:

> My first remembrance of eating wild mushrooms is when I must have been 12 or 13 years old . . . In the early spring our family would have a Sunday afternoon outing to Spring Creek or Rock Creek in eastern Oregon and we would gather morels easily and in quantity that we would share with our friends. Gunny sacks would be half filled by each of us in a short time. My mother cooked them in several ways and I think my favorite way was when she would split them lengthwise and sauté them in butter, adding some heavy cream and serving them atop homemade bread toast. There was a milk-based soup too that I remember in which morels were cut into rings and added prior to serving. With scrambled eggs they were tasty, too. Morels were no big thing—it was like going huckleberrying or wild strawberrying. Something you did, in season.

Should you be so lucky as to gather more morels than you can eat at once, how should you preserve the rest? Drying is a good option because it intensifies the flavor. This is true for other mushrooms, including the king bolete (*Boletus edulis*), but is decidedly not true for many others, which lose flavor and become mushy on rehydrating. How morels are dried does make a difference. The best morel dish I ever tasted was served to me in a restaurant in Germany's Black Forest. When I asked the waiter why they were so tasty, he told me that they had been imported from somewhere in the

Middle East, where they had been dried over the smoke from camel dung!

Harvesting morels is a serious business. Morel hunters treasure the special collecting sites where good specimens are found year after year. The story is told of an elderly European lady who, on her deathbed, called her niece home from America so that she could whisper in her ear the location of her private morel haunt. This preoccupation with secrecy seems quite typical of morel hunters. A gentle Boston mushroom lover, the late Rollo Leach, captured the instinct of the hunter exactly:

> He who secrets reveals
> And nothing conceals
> Somehow never tells
> Where grow morels.

The king bolete (*Boletus edulis*) is king of the kitchen among mushroomers in the United States. To mushroom hunters in eastern North America, stories of the western king bolete bring tears to the eyes. First of all, these mushrooms are much more common in the West Coast and the Rockies than in the East. Second, in the West king boletes tend to be free of "worms" (fly maggots, actually), whereas in the East you have to be quick if you are to gather any specimens before they become riddled with unsightly tunnels. I had my moment of triumph when I picked five large specimens in pristine condition in a pine forest in southeastern Massachusetts, but this is rare. The East makes up for this disadvantage in part by having a variety of other edible boletes, many of which seem less affected by mushroom flies and their larvae.

Perhaps the most commonly picked mushrooms in both North America and Europe are the chanterelles (*Cantharellus cibarius*). Chanterelles are often found in large amounts, and on a good day they can be gathered by the peck in the East and by the bushel in the Rockies and the West. The West has the edge when it comes to chanterelle hunting in other regards as well. Western specimens tend to be much larger, up to two pounds in weight, whereas in the East one seldom finds specimens of more than a few ounces. And western chanterelles, like western king boletes, are not as readily infested by maggots. Chanterelles are trumpet-shaped (hence the name, derived from the Greek *kantharos*, "vase") and yellow-orange in color. They can be distinguished from other truly gilled mushrooms because their gills have a rounded rather than a sharp edge. In older specimens especially, the gills are forked and blunt-edged and look like folds in the body of the mushroom.

Opinions vary about where to rank chanterelles in terms of taste. Almost everyone likes them, but some list them among the very best while others are less enthusiastic. I enjoy their robust flavor, which is far removed from that of all other mushrooms, as well as their slightly chewy consistency. Chanterelles are said to have an apricot flavor, an inadequate description but not one that I could improve. These mushrooms can stand alone as a side dish or can be used as ingredients in a complex recipe without being overwhelmed. Chanterelles lose flavor and consistency on drying, let alone canning, and are best eaten fresh. Once lightly sautéed, they can be kept frozen for a few months.

The West and the East are more alike when it comes to a number of mushrooms that are appreciated but do not necessarily send mushroomers into a state of ecstasy. Some of these species are truly cosmopolitan and are found in abundance in most regions of North America. The oyster mushroom (*Pleurotus ostreatus*) is one such. It is found on diseased tree trunks and old stumps and, in cultivated form, on supermarket shelves. When picked during cold weather, the wild ones taste much better than their domesticated counterparts. Depending on the ambient temperature, the caps arising from the same mycelium vary in color from pure white to light brown to a lead-gray hue. The colder it is, the darker the color of the caps. This species is one of the few edibles that can be found throughout the year, including during a New England winter thaw.

Oyster mushrooms grow out sideways from a tree trunk, with the caps sticking out of lateral stems. They are found in tiers, sometimes in profusion, and can cover a whole tree trunk. It is not clear to me how oyster mushrooms got their name, although they have a vaguely fishy taste and the shape of scallop shells. This is a versatile species that can be prepared in many ways. Margaret Lewis, for many years the guiding light of Boston mushroomers, enjoyed great success substituting this mushroom for scallops in a recipe for Coquilles St. Jacques.

Also cosmopolitan is the honey mushroom (*Armillaria mellea*), often found in large bunches at the base of seemingly healthy as well as dying trees. The name does not refer to the taste but to the color of the caps, although some varieties (different species, according to some taxonomists) are lighter and others browner than the color of honey. Honey mushrooms grow in clusters, have stems that taper downward with a pith-like interior and a flaring ring, and have dark scales on the top of the cap. This is a parasitic fungus that causes damage to many kinds of trees. Prodigious clumps of dozens of individuals can be found, although this mushroom also grows in solitude or in small groups. Several poisonous mushrooms superficially resemble the honey mushroom. They are yet another reminder that every mushroomer should be proficient in identification before eating gilled mushrooms that grow on wood.

Here are some notes on gathering and using honey mushrooms by Al Ferry, a consummate mushroom cook.

Sliced, and cooked in some butter, or steamed, the caps can be kept refrigerated for several days, longer if reheated; or frozen for months. Collected after a rain the larger caps may be damp, almost soggy. These can still be made useful by steaming. Sautéed, armillaria provide a natural thickening, somewhat like using fresh okra, and can be used in chowder and soup. Large pieces are good cooked along with steak.

Unless they are very young and fresh, the stems will remain tough and fibrous, never cooking tender. So collect just the caps. Cut off the cap with a knife, then inspect the cut for insect tunnels. If it is clear, throw the cap in your bag. You may see a single centered tunnel, or a few tiny ones; then cut the cap in half and decide. If the first four or five in a clump are infested, go on to the next tree. Smaller caps in a clump, I have found, are not necessarily younger. They are often stunted by insect damage.

Two other tree-dwelling mushrooms are prized edibles, the hen-of-the-woods (*Grifola frondosa*) and the chicken mushroom or sulfur shelf (*Laetiporus sulphureus*). They are both bracket fungi, members of a large group of tough, woody or leathery fungi, the polypores. These two, however, are tender enough to be eaten and, despite their names, taste quite different. Fresh chicken mushrooms have a slightly fibrous consistency, resembling somewhat that of white chicken meat. When older, these mushrooms are stringy and slightly sour in taste. Sautéed in strips until they turn golden, chicken mushrooms can be served as finger food or, after further slow cooking, can be incorporated into a stew. These mushrooms are orange-pink on the top and bright yellow on the underside in one variety and white in another. They grow on the sides of dying deciduous trees as overlaying broad, shelf-like projections with thick edges. The growth may be quite extensive; harvests of ten or twenty pounds are common. Fortunately, they can be kept frozen for several months without loss of flavor.

The hen-of-the-woods has a distinct vegetable flavor, vaguely reminiscent of eggplant. These mushrooms grow in dense, gray-brown fronds at the base of oak trees and sometimes reach staggering sizes. Specimens weighing over twenty pounds are found often, and seventy-pound giants have been reported. This mushroom gets its common name from its feathery look, but were it not for its drab color I would liken it to a giant flower, perhaps a colossal peony. I once found a hardy specimen of hen-of-the-woods for sale in an open-air food stall in the North End, Boston's Little Italy. I thought this was a good chance to find out what the storekeeper thought he had for sale and perhaps pick up a bit of mushroom folklore. Since I speak Italian, I thought I'd use that first. "Come si chiama questo?" The answer: "Fungo!" I thought I would try again, in English this time. "What do you call this one?" The answer: "Musharoome!"

Many other wild mushrooms generally considered worth picking for the table are listed here. Once past the "big edibles," there is considerable disagreement among mushroom fanciers as to what is worth getting excited about. I have heard people swear by certain clavarias (coral mushrooms), russulas, or some kinds of boletes that others find tasteless. Mushrooms need not be the centerpiece of the menu. Many species can be appreciated because of texture rather than strong taste. An example is the wood ear (*Auricu-*

- Boletes (many species)
- Caesar's mushroom (*Amanita hemibapha* in U.S.)
- Cauliflower mushroom (*Sparassis radicata*) and its relative, *S. crispa*
- Coccora (*Amanita calyptrata*)
- Hedgehog mushrooms (*Hydnum repandum, H. erinaceum, H. coralloides*)
- Inky caps (*Coprinus comatus, C. micaceus, C. atramentarius*)
- Lobster mushroom (*Hypomyces lactifluorum*, parasitizing *Lactarius* or *Russula*)
- Milk caps (*Lactarius* species)
- Puffballs, small and large (*Lycoperdon perlatum, Calvatia booniana, C. craniforme, C. gigantea, C. sculpta*)
- Saddle fungi, or black saddle (*Helvella lacunosa*)
- Tricholomas, or man-on-a-horseback (*Tricholoma flavovirens*)

Other North American Mushrooms Worth Eating

---

laria polytricha) used in Chinese cooking to add a pleasant chewiness to dishes. As Al Ferry points out, slices of the "lesser" boletes or some other not so flavorful mushrooms, such as russulas, can be carefully sautéed until they are crisp and then used as garnish for creamed soups or salads. This is especially useful when there are too few specimens to make a more substantial dish.

In the Boston area, 1994 was a great year for mush-rooms—and the last two weeks of September, superb. Edible mushrooms were unusually abundant and my friends and I were able to fill our baskets repeatedly. I found several fine varieties, although these did not include the best of the region, such as chanterelles and boletes. Some of us went on a gastronomic rampage, eating a different kind of wild mush-room almost every night. I believe that our experience illus-trates what a mushroom fancier can expect in the way of gastronomic delights during a good collecting year. Each dish we prepared could have been made with many other kinds of wild mushrooms equally well.

We found a large number of honey mushrooms that year, but at least on one occasion picking required more care than usual because the largest clumps, growing at the base of diseased oak trees, were surrounded by luxuriant patches of poison ivy. Gathering required complex contortions and near acrobatics, but somehow we managed to get away without adverse effects. Back home, we sautéed the caps and froze most of them. We added some to a simple tomato-based sauce, which we ate over pasta. The dish was okay, not much different from what we could have made using white button mushrooms, although the price was better. The following week we found a more interesting use for these mushrooms. We thawed our frozen cache and mixed it with a handful of rehydrated dried boletes and shiitake and added them to lightly fried chicken pieces plus onions and rosemary. After slow simmering with white wine, the dish produced a heartwarmingly rich sauce that performed extremely well over spaetzle, the German dumplings (egg noodles, rice, or some other starch would have been just as good). Lacking proper training in culinary terms, I can only describe the sauce as "rich mushroomy" in taste, enhanced by the musky flavor of the boletes and the smoky aroma of shiitake. The honey mushrooms were not just an expander for the others but helped impart the strong flavor to the sauce. This way of using mushrooms, as an ingredient in a "natural" sauce, is nearly foolproof—it works equally well in dishes made with chicken, beef, or fish. There is always the danger of over-whelming the flavor of the mushrooms with strong herbs or spices, but if the cook has a light hand, robust kinds such as the honey mushrooms will hold their own.

The same week, my wife found a perfect specimen of a middle-sized puffball, *Calvatia craniformis,* on a lawn of a public park where we used to go for our morning walk. The eight-inch specimen was large for this species, which is seldom more than six inches in diameter. It was perfect for eating, pure white and solid inside. Puffballs have a strong aroma raw, but much of the odor disappears in cooking. This specimen, however, withstood cooking quite well, perhaps because my wife cut it up in fairly small strips, about a half inch in thickness. She sautéed them briefly with onions, parsley, flour, lemon, and parmesan cheese, then we served it to guests as an appetizer on toast triangles. The taste was, once again, richly "mushroomy," with a consistency that was soft but not sticky. What to do with what was left over? The next week, we placed the mushroom on top of quartered pieces of winter squash and cooked them in the microwave oven. Not only did this elevate the squash from "mere vegetable" to "main course," but the various flavors had blended extremely well and gave us a novel combination of tastes.

On the last weekend of September, we went on a mushroom walk in Harvard Forest, in Petersham, Massachusetts. We found a few specimens of a late fall mushroom, *Hygrophorus fuligineus,* that I wanted to introduce to my wife and were given a few more by one of our hosts. This "waxy cap" does not have a widely accepted common name. It is barely mentioned in modern field guides, although it appears prominently in older ones. To someone collecting in New England, this oversight seems odd because this species is often found under pine trees in the late fall. Waxy caps come in many varieties, some in spectacularly bright red and yellow colors. They have in common a waxy feeling, especially on the gills. Almost unique in this group, *H. fuligineus* has a dark, almost black cap. Several of the edible varieties, including *H. fuligineus,* have an added characteristic: they become covered with a thick, translucent, glutinous layer that becomes so copious that it flows over the cap's margin. This material gives the mushrooms a lovely shiny appearance, but it makes them unpleasant to touch, and it is very hard to get rid of bits of pine needles and other debris from the sticky surface. The slimy material is made up of a polysaccharide that functions as antifreeze and protects this late species from the chill of the northern fall.

We fought the cleaning battle with our specimens and got them ready for cooking. Their glutinous covering made them good candidates for a smooth cream soup. We cut up the mushrooms and sautéed them with onions, celery, and parsley, put them through a blender and added cream. We added only a very small amount of flour, as indeed thickening was provided—okra-like—by the mushrooms themselves. A little white wine, paprika, and chopped chives added to the flavor. We did well, because the soup was rich in taste and it had a pleasant velvety consistency. In general, you cannot go wrong by processing mushrooms in a blender to make soup, as this intensifies the flavor to a remarkable extent. Even relatively insipid species, such as most of the russulas, gain from the process.

Our autumnal mycological feast was not confined to mushrooms growing in the wild. A friend who owns a cranberry plantation about an hour southeast of Boston invited us to witness the unique spectacle of "wet" cranberry harvesting. The fields (misnamed "bogs" when in fact they are dry and sandy most of the time) are flooded, and a mechanical beater is used to dislodge the ripe berries from the plants. The released berries float, turning the ponds a bright, shining vermilion, and are then gathered up from the water. My friend does not confine himself to harvesting cranberries but also grows shiitake on sections of trunks of surplus oak trees from his property. Following the biblical entreaty, he does not harvest all the mushrooms but leaves a "corner" for the poor, in this case defined as those bereft of fresh shiitake.

Because they had not been harvested early, the mushrooms had achieved a larger size than usual, some being six inches across and heavy of flesh. On the advice of several cook-books, we decided to use them as a substitute for meat. We chose a simple recipe for a Hungarian goulash and cooked the mushrooms to perfection (as menus of pretentious restaurants would say). We indeed ended up with a successful dish: the mushrooms had a robust consistency, lacking only the fibrous texture of the beef chunks of a regular goulash. The typical flavor of shiitake was retained, despite a generous amount of paprika.

By far the most common way of cooking wild mushrooms in the western world is to sauté them in butter. Butter is thought to be the perfect medium for cooking many mushrooms, but the drawback is that its flavor overwhelms that of the milder mushrooms. Oil is more neutral than butter, but which kind to use? There are many kinds, and gourmet shops stock them like wines. In search of the oil most appropriate for mushroom cooking, six of us got together one winter evening to carry out an experiment. Our mushroom-eating experience was varied and so, we hoped, were our tastes. The group included a cook, an ex-monk, a computer expert, a student of comparative religion, an artist, and myself, a microbiologist. We used both white button mushrooms and shiitake, since they contrast in taste and consistency and perhaps would react differently to various oils. (This turned out not to be the case—we reached the same conclusions with both kinds of mushrooms.) The mushrooms were sautéed separately in eleven different vegetable oils (apricot kernel, canola, grape seed, hazelnut, extra virgin olive, peanut, pecan, rice bran, safflower, sesame [white], and walnut).

We wrote our comments on a score card, careful not to let the others see them. As the evening progressed, we guarded our comments with a solemnity seen at the poker table. Our discussions were animated but we arrived at considerable agreement. The oils were unlike one another and the differences sufficiently distinguishable to evoke similar reactions. Olive oil, to no one's surprise, came out "rich and evocative," whereas walnut oil made everything taste fishy. Peanut oil was okay. Rice bran oil had a "toasty flavor" that seemed to bring out the flavor of the mushrooms very well. Another of the lesser-known oils, from apricot kernels, earned the verdict of having a strong fruity taste and being "light but rich." None of us had used grape seed oil before, but we all agreed that it was peppery and had a potato-like flavor. Other oils (canola, pecan, safflower, light sesame) earned a nondescript label. Hazelnut oil occupied one extreme in taste, for it made the mushrooms taste strongly of hazelnut (not a good idea). Our main conclusion was that apricot kernel, olive, peanut, and rice bran oils are particularly well suited for keeping alive the taste of the mushrooms. If that is your goal, use these oils in lieu of butter.

For the imaginative cook, wild mushrooms are open to much experimentation. I believe that the surface of this topic has barely been scratched and that we know how to take full advantage of a few species only. As long as you are not too greedy, you can be reasonably sure that, if not today, then next week you can return from the woods or fields with a collection of specimens worth eating. Why not invent new, perhaps wondrous combinations?

The life of a mushroom lover is rich in contrasts. I remember a dinner party I gave one fall evening for eight close friends. It was as elegant as I could make it: I used my best linen, china, and silver, and the dining room was lit by candlelight. The soup was made with black trumpets and oyster mushrooms I had collected a few days before. Inevitably, the conversation turned to wild mushroom hunting. My skills were affirmed and my bonds with nature envied. Those I had not yet taken on a mushroom hunt extracted the promise that I would do so soon. The main dish was a roasted chicken bedecked with porcini and adorned with a colorful spray of vegetables. My thoughts wandered. I remembered walking through the woods a few days before, dressed in faded blue jeans and an old jacket, looking unkempt and rustic. I mused about the contrast between my woodsy persona and the elegance with which we were surrounded that night. I felt that the natural environment was happily encroaching upon the formal dining room. Obviously, life with mushrooms offers a lot of latitude in style.

No space is wasted underneath the cap of the mushroom. Gills in a variety of forms and colors have evolved to increase the surface area on which spores are produced.

Cortinarius sp.

Gills

Laccaria laccata
(common laccaria)

Daedalea quercina

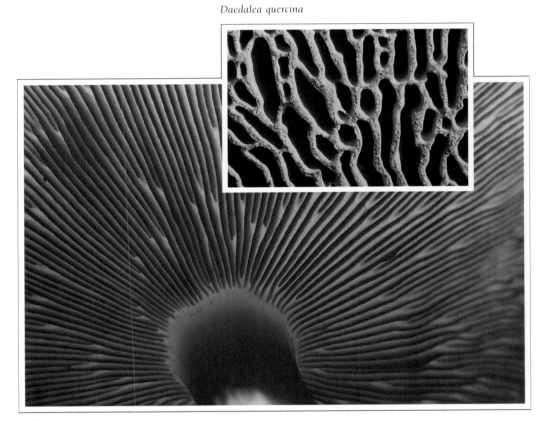

Collybia maculata

Alternatives to Gills

Spores are made not only on gills but in sacs, columns, and open surfaces. Mushrooms take different forms to disperse the spores.

Crucibulum vulgare
(bird's nest fungus)

Clavaria fusiformis

Geastrum saccatum (earth star)

Peziza badioconfusa

Color

You don't need to eat a mushroom in order
to enjoy it. Many mushrooms have intriguing
shapes and dramatic colors.

Chlorociboria aeruginascens (blue stain)

Laccaria amethystina

Hygrocybe cantharellus

Helotium citrinum

Poisonous and Hallucinogenic Mushrooms

Mushrooms don't come with warning labels. You have to know what you have in your hands before you put it in your mouth.

Gymnopilus spectabilis (Laughing Jim)

Amanita virosa

Amanita muscaria
(fly agaric)

Mushrooms don't advertise their edibility. What is appealing to the palate may not be appealing to the eye.

Leccinum aurantiacum

Cantharellus cibarius (yellow chanterelle)

Edible Mushrooms

Morchella esculenta (morel)

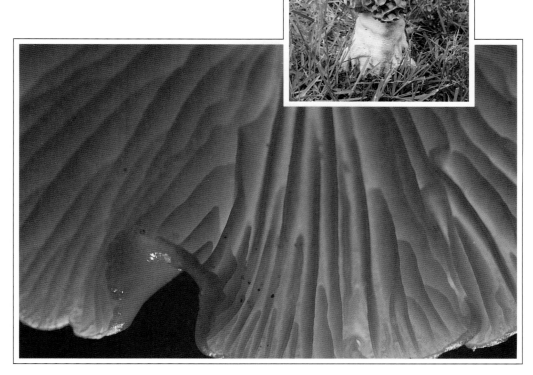

Cantharellus cibarius (yellow chanterelle)

Hypomyces lactifluorum
(lobster mushroom)

Craterellus fallax
(horn of plenty)

Bracket Fungi

Bracket fungi (polypores) are the sturdy ornaments of the forest. Many linger throughout the seasons, and some can be appreciated year after year.

Polyporus squamosus

Pycnoporus cinnabarinus

Trametes versicolor (turkey tail)

Fomes fomentarius

Bracket Fungi

Trametes versicolor (turkey tail)

8

WHITE BUTTONS, SHIITAKE, AND OTHER TAMED SPECIES

If you are told that a dish on the menu of a fancy restaurant has "wild mushrooms" in it, you can be pretty sure that the mushrooms were cultivated. The species likely to be served are shiitake, portobellos, or perhaps oyster mushrooms. These are wild only in the sense that anything other than the store-bought white button is a "wild" mushroom. In time, it is possible that the term "wild mushrooms" will be restored to its original meaning and include only such kinds as boletes and chanterelles that indeed cannot yet be cultivated. Whatever they may be called, many more species can be cultivated than are currently available in North American food stores. Thus, in the near future we may be given the choice of an even greater variety of "wild mushrooms." Around the corner may be such delicacies as morels and even truffles.

The white button mushroom commonly sold in markets is said to have originated in France, where it is known as "champignon de Paris," or plain "champignon." It seems to have been first cultivated in earnest around 1650 in stone quarries around and underneath Paris. (See Figure 8.1.) It was later grown in caves, which has added credence to the belief that mushrooms like to grow in gloomy, shadowy places, although in reality most temperature- and humidity-controlled locations will do. In the United States, the center of commercial cultivation has been the small town of Kennett Square, Pennsylvania, where mushrooms are grown in long, temperature-controlled "mushroom houses." White button mushrooms keep and travel unusually well, especially after treatment with sulfites, and that is the main reason they have become so widely available. They are dung-loving and traditionally have been cultivated on a compost of horse manure and straw. Not to worry, though: as the compost is prepared, fermentation takes place and generates enough heat to kill most bacteria and other fungi.

The common mushrooms are sold as immature buttons, before they have started to make spores in earnest. The reason has more to do with marketing than with taste. As spores are produced, the gills go from a reassuring pink hue to dark brown and eventually to a less appealing black color. It may be the case that with the gills hidden from view, the buttons have a more familiar "vegetable" shape. With age, however, the flavor intensifies—selling the button stage represents yet another triumph of marketing over taste. Luckily, the reverse is now often the case, and different varieties of the white button mushroom are sold in an advanced state of maturity under exotic-sounding names ("porto-

bello," "crimini"). These are simply different varieties of the same species of mushroom.

Consumption of mushrooms in North America shot up after World War II, but the growth was confined entirely to the white button. By now most people here and elsewhere have run into other cultivated kinds, such as portobellos and crimini (varieties of *Agaricus bisporus*), shiitake (*Lentinus edodes*), dry or fresh, oyster mushrooms (*Pleurotus ostreatus*),

8.1
Mushroom cultivation in caves near Paris in 1868.

and enoki (*Flammulina velutipes*). U.S. sales of such cultivated "specialty" mushrooms reached 40 million pounds in
the 1994–95 season, three-quarters of it portobellos. Shiitake
accounted for 5.5 million pounds. For comparison, the 1994
U.S. production of the white button mushroom was over 780
million pounds. One may also find in fancy food markets
genuinely wild, uncultivated mushrooms, such as chanterelles, boletes, or morels. They usually cost a pretty penny:
dry porcini (*Boletus edulis*) and morels are sold in specialty
stores in little cellophane bags for several dollars per quar-
ter-ounce.

Many more kinds of mushrooms than these are cultivated
in the Far East. Among those grown in enormous amounts
is a whitish round mushroom found in Chinese dishes called
the "paddy straw mushroom" (*Volvariella volvacea* or a close
kin). The Latin name of both genus and species indicates
this mushroom, especially when young, is entirely enclosed
in a membrane, called a "volva." Specimens are harvested in
the "egg" stage, when the volva is still intact and the mushrooms resemble a quail's egg. Next time you order a dish
with these mushrooms in a Chinese restaurant, why not try
an on-the-spot study of mushroom anatomy? You won't be
able to do it with chopsticks, though, so swallow your pride
and ask for knife and fork. Cut across one of the "eggs," and
inside it you will see a baby mushroom with a perfectly
formed cap, a stem, and gills. The canned form, which is
how we are likely to encounter this mushroom, is mild
tasting. Fresh specimens have a stronger and richer flavor,

and Paul Stamets, an expert in mushroom cultivation and the author of several authoritative books on the subject (see "Resources"), has pronounced this species one of the best of all edible mushrooms.

Until recently, it was believed that morels could not be cultivated artificially, but a remarkable breakthrough was made in the 1980s. A patent for the artificial cultivation of morels was granted in 1986 to a young California graduate student, Ronald Ower, and two collaborators from Michigan. Ower had discovered that by simulating the changes in moisture and availability of nutrients that occur naturally throughout the seasons, he could obtain a high—although perhaps not wholly predictable—yield of morels. What he did sounds simple enough. He grew the morel mycelium on moistened wheat to which he had added nutrients, and eventually, as in the wild, the fungal growth developed into a hard mass of filaments filled with internal nutrients, called a "sclerotium." At this point, the external nutrients are removed and water is added. The fungal mass is now so loaded with nutrients, such as fats and carbohydrates, that it can support the development of morels without much further feeding. (This is a simplified description of the procedure, which in reality requires attention to a number of other details.) The researchers claimed in the patent application that they could get 25 to 500 morels per square meter of growth surface. Using the upper figure, I fantasized that if New York's Central Park were used for cultivating morels, the yield would be close to 200 specimens per man, woman, and child of that city every few months! The patent rights for morel cultivation have wandered around somewhat. First they became the property of a national pizza company. Most of the cultivated morels are sold to restaurants and specialty food stores. The commercial production has been small and cultivated morels have yet to make their way to the grocery store, although the day when we may simply add this delicacy to our weekly shopping list may be near.

For "couch potatoes" who don't want to go out and hunt for wild mushrooms in the woods, there are several varieties that can be grown at home. In my mind that's like raising grouse to shoot it, but there is no law against it. The easiest way to start a farm in the comfort of your own home is to get a kit that includes the substrate on which to grow the mushrooms and the mycelium (known as "spawn") to inoculate it. You can choose, among others, white button mushrooms, shiitake, oyster mushrooms, wine cap mushrooms *(Stropharia rugoso-annulata)*, and, more recently, morels. Perhaps the easiest ones to cultivate are the oyster mushrooms, which are nearly omnivorous and will grow on wet straw, sawdust, or shredded newspapers as long as they are mixed with oatmeal or the like. There are even kits for outdoor cultivation, both on the ground and on stumps. For the more adventurous, it is possible to start from scratch. The usual way to begin is by culturing bits of wild mushroom on agar in Petri dishes. This requires sterile techniques and some familiarity with microbiological methods. The resulting spawn is then carefully transferred to suitable substrates and, if one is lucky and has been able to avoid contamination by molds, mushrooms will eventually appear.

The white button mushroom can be cultivated with relative ease in any place as long as the temperature is kept steady at around 55°F. An article from early in this century in *The Garden Magazine* (October 1906) by Loise Shaw from New Jersey tells of using a Ping-Pong table as a surface for constructing a mushroom bed. The sides of the table were boarded up to a height of fifteen inches and the bed made from a compost of horse manure. Three dollars sufficed to buy fifteen wheelbarrowfuls from a nearby livery and for two dollars, sufficient spawn was purchased to inoculate the whole table. The first crop appeared two months after the bed had been inoculated, and a total of forty-eight and a half pounds of mushrooms was gathered.

In forests and fields there are many wild counterparts to the domesticated varieties found in supermarkets. The white buttons are close relatives of the field mushrooms, and the whitish oyster mushrooms in the store belong to the same species as those that grow on trees and stumps. The cultivated wood ears (*Auricularia polytricha*), often found in Chinese soups, have a wild relative of the same shape and flavor, although this species is not often found on this continent. The North American woods do not have a natural counterpart to shiitake, which must be bought in the store or cultivated by the more enterprising mushroom fanciers.

Wild and cultivated mushrooms of the same species often look similar, but in some cases they differ considerably. Most

surprising is the difference in appearance between the cultivated and the wild form of enoki *(Flammulina velutipes)*, a
mushroom that is sometimes found in supermarkets. The
cultivated kind is sold as a white, spaghetti-like bundle, each
individual thread ending in a tiny round, white cap. The outdoor form looks totally different and one would never think
of the two as belonging to the same species. In the wild, this
mushroom is much stouter, usually an inch or so across,
with a brown, shiny cap and a stem that is prominently dark
brown at the base. Domestication has changed this species
beyond recognition. Interestingly, the two forms offer completely different gastronomic experiences. The store-bought
form is meant to be eaten raw on salads and has a grape-like
taste, whereas the wild variety is usually cooked and tastes
more like the white button.

In North America, commercial picking of wild edible
mushrooms used to be carried out on a small scale, usually
for sale to fancy restaurants. Now, however, this business has
taken the larger scope of supplying food markets here as well
as for export abroad. In recent years, foraying for profit has
reached epidemic proportions in the Pacific Northwest. The
most often harvested species, at least in dollar value, are
chanterelles and matsutake. Most of the chanterelles make
their way to Europe to be canned into something resembling
soggy yellow cardboard. Thousands of pounds of wild mushrooms are gathered by licensed and unlicensed pickers, who
deliver their harvest to processing stations. The 1990 figure
for the state of Washington alone is nearly a half-million

$$\langle\!\!\!\text{\char"20DD}\!\!\!\rangle\!-\!CH\!=\!CH\!-\!C\!-\!OCH_3$$

8.2
Methyl cinnamate, the chemical that gives matsutake mushrooms their flavor.

pounds for the ten species most often picked, but this is an underestimate because only a portion of the harvest is reported. The 1990 average wholesale price in Washington was $2.59 per pound; an active collector could make several hundred dollars in one day, depending on the species picked. Of course, the season for peak collecting is usually short, limited to a few weeks. To maximize their income, some professional pickers follow the harvest, starting in northern California and going beyond the Canadian border.

A close relative of matsutake (*Tricholoma matsutake*), the most prized of all mushrooms in Japan, is collected in large quantities in British Columbia, Washington, and Oregon in fir and pine forests. In 1989 the wholesale price in Washington was $14 per pound. The American matsutake (*Tricholoma magnivelare*) is transported fresh by air to Japan, where it fetches stunning prices, sometimes over $200 dollars per specimen. Conventional wisdom in Japan has it that the American and Korean varieties are inferior to the native Japanese form, but the sale of the imported varieties is substantial nevertheless. The "cinnamony" smell of matsutake is due to a relatively simple chemical, methyl cinnamate (Figure 8.2). This is only one of various substances responsible for the complex aroma of cinnamon, but it alone is responsible for the "matsutake" odor. Adding methyl cinnamate to otherwise bland mushrooms elevates them in aroma to the exalted ranks of matsutake. The extent to which this will be done commercially remains to be seen. In addition, there are reports that matsutake has been cultivated in the laboratory, but apparently it has not yet been raised commercially.

The harvesting of edible mushrooms in the wild is done on such a large scale that it is feared it may interfere with the pleasures of amateur collectors. A controversy is raging between those who want to reserve the woods for amateur collecting and those who welcome the opportunity to harvest a renewable resource at potentially little cost to the environment. Vehement as the argument has been, it has been hyped up by the press, which was quick to report a murder among mushroom pickers supposedly vying for the same collecting grounds ("It Seems Some Folks Would Simply Kill for a Nice Morel," *Wall Street Journal,* May 11, 1993). According to later reports, the killing was connected to gambling debts and unrelated to competition for the picking territory.

Here is the gist of the argument. Amateur mushroomers complain that as the result of large-scale collecting there are fewer mushrooms to go round and that the environment is disturbed by the careless trampling and raking by pickers as they ransack the woods. The other side says that mushrooms are a natural and renewable harvest that provides a livelihood for those willing to work hard. The state of Washington is the only one that requires a license for both picker and dealer ($75 and $375 yearly, respectively) and attempts to regulate the industry, although others are working on it. When officials of the U.S. Department of Agriculture looked into the matter and asked state agricultural agents for information, they got from Texas a Texas kind of answer: on official stationery, it was explained that the harvest of mushrooms over six feet tall or three feet wide is prohibited there. In addition to the obvious issues concerning commercial picking, it is possible that the value of the harvested mushrooms in some forests may exceed that of the timber. If mushroom harvests are taken into account, perhaps the woods will be spared from lumbering operations. In 1992, the Oregon Natural Resources Council filed a court appeal to protect a forest in the Crescent Ranger District from logging, pointing out that the export value of matsutake alone, $15 million for 1989, was greater than that of the lumber.

Does overpicking in fact limit future yields? There is little reliable information on this point, and estimates of the effect of widespread harvesting on future crops are based largely on anecdotal information. In principle, collecting mushrooms is likely to be less detrimental than is usually imagined. Mushrooms, being the fruiting bodies of certain types of fungus, are something like the fruit of a tree or the berry on a bush: the individual organism is little affected by harvesting the fruit. In a carefully designed study on the effects of picking chanterelles in Oregon, the researcher Lorelei

Norvell reached the tentative conclusion that no real effect was evident. She cautions, however, that this matter may not be so simple and that more research is needed before the point can be settled. Meanwhile, the argument goes on. California and other states have enacted the same blanket edicts for both amateur and commercial collectors and have recently curtailed picking on public lands, much to the annoyance of everyone involved in mushrooming. It should be recognized that the artificial cultivation of mushrooms, like other forms of agriculture, also entails an environmentally relevant cost of materials and energy. In a way, the argument regarding mushroom picking seems misplaced. The overriding factor in reducing the growth and availability of mushrooms is the destruction of natural habitats. That should be the biggest concern.

In Japan a large industry, backed by research organizations, has been established in order to satisfy the enormous desire of people for mushrooms—or *kinoko,* as they are called. Only there can one find a Mushroom Park, consisting of a research institute, a museum, a shrine, a hotel, and a restaurant, all devoted to mushrooms. It is in the town of Kiryu in Gunma Prefecture about two hours north of Tokyo, where the Kanto plain meets the mountains. Kiryu was chosen for the Mushroom Park both for the beauty of its surroundings and because it is the native place of Dr. Kisaku Mori, who discovered the modern method of growing shiitake on oak logs.

Before the 1930s, shiitake growing was widely practiced in East Asia, but it was a hit-or-miss proposition. Logs were seeded superficially with infected wood, a method that was neither reliable nor efficient. Consequently, mushroom farming was not a source of dependable income. As an agricultural student in Kyoto University, Mori experimented with the production of mycelium-impregnated wooden dowels, which would be inserted into holes drilled in the logs. This procedure has since been perfected, and good, steady yields of high-grade varieties of shiitake are the result. It has made it possible to cultivate shiitake as a major agricultural crop in Japan, China, and other East Asian countries, and it is the method widely used by shiitake growers in this country. Mori's success led him in 1936 to found the Kiryu Institute devoted to research in mushroom cultivation. In later years, a commercial offshoot was added, and, more recently, the hotel and restaurant.

It was my good fortune to visit the Mori complex in February 1992, at a time when spring began to make itself felt in that part of Japan. My host, Professor Kazuo Nagai, and I began the visit with a most satisfying meal at the restaurant. It included five kinds of mushrooms—shiitake (*Lentinus edodes*), enoki (*Flammulina velutipes*), maetake (or "hen-of-the-woods," *Grifola frondosa*), nameko (*Pholiota glutinosa*), and honshimeji (*Lyophyllum decastes*)—in a "nabe"-type, "cook-it-yourself" dish. Afterward, the director, Dr. Takeshi Nakazawa, graciously showed us around the installation. He told us of a party of eager American amateur mycologists who visited him a few years ago. The visitors had come to Japan on a mushroom tour, one of the excursions organized yearly to the four corners of the world. This group was led by Gary Lincoff, a well-known teacher and researcher of mushrooms from New York City.

We were told by Dr. Nakazawa that perhaps the most important recent development in mushroom cultivation in Japan was the increased availability and popularity of the hen-of-the-woods mushroom (*Grifola frondosa*). This species, prized for a long time by the Japanese, is known as "maetake," or "dance mushroom," because, according to one dubious story, of the joyful dance performed by anyone who finds it. Currently, this is one of the common mushrooms available in Japanese supermarkets and food stores; it is sold at a reasonable price and is vying with shiitake for number one in popularity (matsutake is too expensive for most people). We are now able to buy this delectable mushroom in North American supermarkets.

Dr. Nakazawa wistfully related that shiitake cultivation is shifting from the log procedure for which his institution is famous to a method that uses compacted sawdust blocks. The research being conducted at Kiryu is targeted toward improving strains for cultivation. Much effort is expended on creating hybrid varieties by a technique for cell mating called "protoplast fusion," which consists of denuding mycelial filaments of their cell walls and inducing different types of filaments to fuse. This technique is not in itself demanding, but, given the complexities of fungal genetics

8.3
A shrine on the
grounds of
the Mushroom
Research Institute
and Mushroom
Park, Kiryu,
Japan.

(see Chapter 3), stable hybrid strains are not readily obtained. This research is being vigorously pursued both at the Kiryu laboratories and other agricultural research institutes in Japan.

My visit included spending a few minutes in meditation at the mushroom shrine in the woods and then going to the hotel perched up on a hill (see Figure 8.3). In the lobby of the hotel is a bronze statue of Kannon, a "bodhisattva," a member of the Buddhist pantheon, surrounded by sprouting shiitakes (is there in the Christian world a monument to "Our Lady of the Mushrooms?"). The gift shop offered a collection of mushroom memorabilia, which included field guides, dried and canned mushrooms, mushroom books for children and assorted knickknacks, such as mushroom key rings and bottle openers. The Kiryu museum had on display a large collection of wild mushrooms, pickled in jars and dry, plus posters and photographs of Japanese edible and nonedible species. The emphasis is clearly on education.

A trip to Japan offers the mushroom lover a chance to immerse onself in mushroom culture and to commune with people who have an ingrained affinity for them. The visit would not be complete without a trip to Kiryu. Where else can one make a mycological pilgrimage?

9

TRUFFLES, STINKHORNS,
AND CORN SMUT

Some years ago I was sitting proudly at the head table in a fine restaurant in Milan. The occasion was a scientific meeting in honor of the Italian molecular biologist and humanitarian Luigi Gorini. I was trying not to eat too much of the magnificent food, so I would be ready for my after-dinner presentation. Then came the risotto. To hell with caution. The dish was too good. If that were not enough, the waiter came by holding in a napkin something that looked like a potato and carefully shaved off slivers of the stuff to top the risotto. I was very excited when told what these were: white truffles! Not being shy when it comes to such matters, I asked him for some more slices. Only later did I find out that, in view of the cost, the number of slices of *Tuber magnatum* is counted, and my request had overstepped the bounds of proper behavior!

Most people know that truffles cost a fortune, that they grow in the ground under oak trees, and that they are rooted out by French pigs. The cognoscenti may even know that the truffle-hunting porcine is a female. Truffles emit a chemical similar to a sex hormone to which sows are attracted. (Interestingly, truffles have been regarded since Greek and Roman antiquity as aphrodisiacs for humans as well, but there is little evidence that they have the desired effect.) Since pigs are not the most mobile of animals, truffle collectors convey them to likely sites in a wheelbarrow. Dogs are used in Italy for the same purpose, but they must be trained to go after the smell of truffles. This is an expensive investment, because real truffles must be used in the training period. Considering the fact that the white truffles in Piedmont are sold for as much as $500–1,000 per pound in the local stores, a good truffling hound is worth a fortune. In France, much of the truffle collecting is done directly by humans. Some people have a particularly fine nose and can detect the characteristic odor even if a truffle is a foot underground. Others look for characteristic cracks in the soil under oaks or for swarms of tiny "truffle flies" hovering over the site of buried specimens. There has even been an attempt to make a hand-held device to detect the truffle aroma above ground. This is a portable gas chromatography apparatus, but its usefulness has yet to be established.

Comparing the gustatory qualities of the black French truffles and the white Italian ones will spark a debate, with staunch partisanship on both sides. Contributing to regional pride may be that both kinds are harvested in relatively small areas. An Italian truffle expert, Enza Cavallero, has been quoted as saying that "the difference between the truffles of

Italy and France is the difference between angels and nuns." I could be persuaded to volunteer for a meal where the merits of the two kinds of truffles were to be adjudicated.

Truffles grow two to twelve inches below the surface of chalky soils. Both the black and the white truffles may reach three or four inches in diameter, but a specimen of this size would be a phenomenal find. Truffles are roughly potato shaped. The black truffle has a blackish, rough coat and an interior that appears marbled owing to irregular black and white striations. The dark "veins" consist of spores that can barely be seen with a strong magnifying glass.

Most of the legendary black truffles (*Tuber melanosporum*) harvested in the Perigord region in the southwest of France do not grow wild but are cultivated in association with oaks. There are several techniques for growing the "black diamond," including planting acorns that have been in contact with soil containing the truffle mycelium. Some people bury pieces of a truffle or use artificially cultivated spawn to inoculate oak seedlings. There are growers who sell not only the truffles themselves but also inoculated oak seedlings. In all cases, it takes several years for the trees to grow and the truffles to develop. The close association, or symbiotic relationship, between truffles and their host trees is called "mycorrhiza," whereby the fungal filaments invade the roots of the trees and help them absorb nutrients. This subject is treated in greater detail in Chapter 12. Not only oaks but also hazel trees and even acacias (in Africa) have been used for truffle cultivation.

These techniques have been tried recently in places other than France, including the United States and New Zealand. An experiment was carried out in 1980 by Franklin Garland, who inoculated oak and filbert seedlings with black truffles and planted them in his North Carolina orchard. It took thirteen years, but eventually he succeeded in obtaining a small but significant crop. Halfway through this period, the "truffle hound" he had trained died and he had to invest in training a new one. Patience, then, is the hallmark of truffle cultivation. There is every reason for trying to cultivate truffles in various parts of the world, as the French production has diminished in recent years, apparently because there is less planting of new oak trees. French truffles have risen in price accordingly, and they can fetch almost as much as the white truffles, depending on the locale. No wonder that in some French restaurants truffles are kept under lock and key!

Cultivating truffles is expensive and time-consuming, but growing the mycelium alone is relatively less demanding. The cultivated mycelium yields an edible product with some of the flavor of the fruiting body itself. "Truffle oil," impregnated with the flavor of truffle mycelium, can be purchased in specialty stores. One way to extend the precious truffle is to store them in a closed container together with fresh eggs. The omelet made with these eggs acquires a fine truffle flavor. Truffles are often sold in tiny cans, where they retain only a pale hint of their true aroma.

There are several species of wild truffle native to North America, at least one of which is highly desirable. It is known as the Oregon white truffle (*Tuber gibbosum*) and resembles in flavor the Italian white truffle. Its aroma is said to resemble that of a mixture of garlic and pungent cheese. This species has yielded to cultivation methods similar to those used with the French truffle, but seedlings of the Douglas fir are used as hosts. Truffle hunting is a specialized area of mushrooming, and the interested reader should contact the North American Truffling Society for more information (see "Resources").

In North Africa and the Middle East, places not known for mushrooms, there are yet other kinds of highly edible truffles whose widespread use is not often recognized elsewhere. The most widely collected one is the "desert truffle," or *terfez* (*Terfezia leonis* and others), which grows abundantly in sandy soils and is frequently sold in local markets throughout the Golden Crescent, from Morocco to Iraq. I had the opportunity to try desert truffles on a trip to Israel. My hosts were of Moroccan origin, but I didn't know at the time if these mushrooms are found as far west as Morocco. I took a chance and asked if they knew desert truffles, to which they enthusiastically replied, "Ah, terfez!" Not only that, they took from the freezer a pot containing two or three pounds of what in color and texture looked very much like quartered, medium-sized potatoes. In a dish prepared with a light tomato sauce, the distinctively aromatic flavor of the terfez stood out. The taste was delicate, not pungent like that of the French or Italian truffles, but I found it most pleasant nonetheless.

I had previously read that people in Iraq used to look for terfez by rubbing their big toe over the dry ground to find areas that felt harder than the rest. What makes this digital exploration possible is that terfez grow near the surface of the ground and not at the depths of the "classic," forest-dwelling truffles. As my Israeli hosts told me, the technique is used in Morocco as well. Unlike the French or Italian truffles, terfez are collected and eaten in large amounts over a vast region. It seems unfair that this important food source gets little play in most western books and articles about edible mushrooms.

Humans, pigs, and dogs are not the only animals that are attracted to truffles. So are subterranean small mammals and insects, which play an essential role in the dispersal of the truffles. Truffle spores pass through the digestive tracts of these animals intact and are thus deposited wherever the animal goes. The reason truffles smell, then, is to attract animals and get eaten. In at least one case, the relationship is very intimate: truffles serve as the principal foodstuff for a particular California field mouse, the red-backed vole.

Truffles are fundamentally different from most other mushrooms. Truffles, morels, and the cup fungi belong to one of the main branches of the fungal kingdom, the as-comycetes. The name is taken from the *ascus*, the sac that

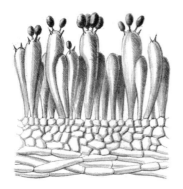 

9.1
The two main modes of producing sexual spores: In the Basidiomycetes (*left*), which include most mushrooms, the spores bud from cells called "basidia." The Ascomycetes (*right*), which include morels and most yeasts, encase their spores in sacs, each of which is called an "ascus."

contains the spores produced by the fungus. Unbeknownst to most mushroom hunters, ascomycetes are the largest group of fungi and include most of the microscopic molds and yeasts. The majority of the fungi that cause human disease, such as *Candida, Aspergillus,* and the agents of histoplasmosis and coccidioidomycosis (Valley fever), also belong to this group. In many ascomycetes, the spores are even more forcibly discharged into space than are the spores produced by gilled fungi. This discharge is sometimes so violent that you can actually hear it if you place a cup fungus against your ear. Most of the mushrooms, in contrast, belong to the basidiomycete branch of the fungus kingdom. Their sexual spores are made on cells called *basidia.* Basidiospores are not encased in a sac but emerge freely from the basidia. (See Figure 9.1.)

Truffles provide only one example of the use of odor to

attract animals for the purpose of dispersing spores throughout the environment. A similar strategy is used by a group
of mushrooms called the stinkhorns, although in most other
regards these mushrooms could not be more different from
the truffles. The stinkhorns are basidiomycetes and are relatives of the puffballs, but they differ from the puffballs in
that they carry their spores on their outside surface, on a
stalked cap covered with slime. As these fungi make their
spores, they begin to stink to high heaven, emitting an odor
that flies find as attractive as that of rotting flesh. To the flies
these mushrooms are good places for laying eggs. When the
flies land on the cap of a stinkhorn, they ingest the spores
or pick them up on their legs and other body parts and carry
them to distant sites. Human beings tend to be repelled by
the stench, which to us resembles that of carrion (the of-
fending chemicals include methyl mercaptan and hydrogen
sulfide).

People react strongly to stinkhorns not just on account of
their smell. Some members of this group have a startling
resemblance to the human penis, others to the penis of the
dog. Surely this combination of attributes has helped this
group of funguses, the Phallales, earn a prominent, if dis-
reputable, place in the history of mushrooms and human
affairs. Hardly any type of fungus has produced a greater
reaction than the stinkhorns. Imagine a well-bred person of
the Victorian era finding in the woods a phallic-looking
object with a green cap, covered with flies, and emanating
an unspeakable stench. Small wonder the sight inspired

strange beliefs. Stinkhorns have been associated with all manner of misery and mischief. They have been called devil's eggs ("Daemonum ova") and have been blamed for witch-craft, cholera epidemics, and untold other disasters. David Arora makes a good point when he writes: "When stinkhorns are discussed, the language makes a startling and unprece-dented qualitative leap, from monotonous minutiae to half-baked hyperbole, as if the authors were suddenly taking an interest in what they were saying." Here is an example from *Period Piece,* a book about life in Victorian Cambridge by Gwen Raverat, a niece of Charles Darwin:

> In our native woods there grows a kind of toadstool, called in the vernacular The Stinkhorn, though in Latin it bears a grosser name. This name is justified for the fungus can be hunted by the scent alone; and this was Aunt Etty's greatest invention: armed with a basket and a pointed stick, and wearing special hunting cloak and gloves, she would sniff her way round the wood, pausing here and there, her nostrils twitching, when she caught whiff of her prey; then at last, with a deadly pounce, she would fall upon her victim, and then poke his putrid carcass into her basket. At the end of the day's sport, the catch was brought back and burnt in the deepest secrecy on the drawing-room fire, with the doors locked, because of the morals of the maids!

Stinkhorns are mentioned by the earliest western writers, Pliny the Elder among them. The first booklet written on any specific kind of mushroom was in fact about stinkhorns, published in Holland in 1562. Herbals and other botanical works included them faithfully, as can be seen in the accompanying figure, which the author plagiarized from a previous book by Clusius but printed it upside down (Figure 9.2). Gerard, the author of the herbal, gives the stinkhorn of the figure the Latin name *Fungus virilis penis erecta forma.* The common North American stinkhorn sports the name of *Phallus impudicus.* Though generally viewed with distaste, stinkhorns have had some good press as well: they have been reputed to cure epilepsy, gout, and rheumatism and, not surprisingly, added to aphrodisiac potions for both man and beast.

Stinkhorns are found in a variety of habitats, including sandy and cultivated soils. They may be abundant and can be detected by the dozens in mulched areas or on piles of old wood chips. When ripe, they announce their presence because their fragrance can be noticed a hundred feet away. Stinkhorns grow out of rounded "eggs" the size of a Ping-Pong ball partially buried in the ground. At this stage, they are totally encased in a whitish membrane and, by looks alone, may be confused with puffballs. However, these spherical bodies have a rubbery feel, since the membrane floats over a layer of jelly. When this egg stage of the stinkhorns is cut through, it reveals a rather complex structure unlike the uniform texture of a puffball. In cross-section, the egg reveals a greenish cap and the beginnings of the hollow stalk, all encased in a gelatinous mass. At this point, stinkhorns do not stink and can be handled with impunity.

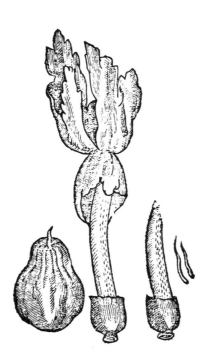

9.2
In a famous
seventeenth-century
herbal by Gerard,
this drawing of a
stinkhorn was copied
from a previous work
by Clusius, but it was
reproduced upside
down.

Stinkhorn eggs mature rapidly into full-fledged specimens eight to ten inches tall. The stem soon breaks the enveloping membrane as it enlarges at a speed of about four to six inches an hour. (See Figure 9.3.) The expansion of the stem is so forceful that if a stinkhorn egg is allowed to mature in a closed jar, it may break the glass. Once the stinkhorn has expanded, spores are formed and cover the cap with a greenish slime. Flying insects such as bluebottles and other flies

9.3
The stinkhorn (*Phallus impudicus*) develops in a few hours from the "egg" stage to the mal-odorous mature form that attracts insects.

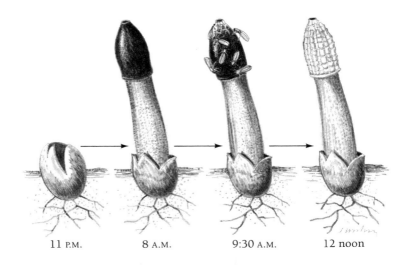

11 P.M.   8 A.M.   9:30 A.M.   12 noon

alight on the cap, the spores are picked up, and soon the cap becomes naked and colorless. Having served its purpose as a spore disperser, *Phallus impudicus* now becomes flaccid and falls to the ground, eventually to become a shapeless mass.

Surprisingly, stinkhorns, once cleaned of the slime on the cap, are good to eat. Friends of mine and I once ventured to eat stinkhorn eggs, much to the astonishment of bystanders who thought that we were indulging in bizarre extremes of mycogastronomy. We tried a few ways to prepare them, including "hard boiling" them, but eventually we settled on sautéing them in oil. The result was a flavorful dish with a subtle, radish-like flavor. The part of the egg destined to become the stem was particularly crunchy, resembling puffed rice cakes. For a banquet of the Boston Mycological Club, one of the experts in collecting edible mushrooms, Ben Maleson, brought a very fancy hors d'oeuvre consisting of stinkhorn stems stuffed with a tasty cream filling and cut into rings. No more refined dish has passed my lips.

It is in China that stinkhorns rise to their deserved appreciation. They are thought to be the most exquisite of mushroom delicacies. Both washed mature specimens and the eggs are eaten there. Mature dried specimens cleaned of the smelly spore slime have sold for $500 a pound in Hong Kong. Stinkhorns are found in profusion in the bamboo forests of southwestern China, where the giant pandas live. Because of their popularity, attempts have been made to cultivate these fungi, but I venture the guess that it will be some time before they are offered for sale in our local markets. Meanwhile, we must make do with the specimens we find in the wild. Stinkhorns do not deserve the reputation they have received. Truffles need no champions, but stinkhorns definitely do.

Nothing could be more wrong than to think that all edible mushrooms resemble the white buttons of the supermarket. In fact, many are so different in appearance that they effectively conceal their kinship to the agarics found in the produce department. To help dispel the "agaricentric" view of the world, I will present a few examples of alternatives. I will start with corn smut. To the consternation of corn farmers, 5–10 percent of all ears of maize in the United States used to become infected by the corn smut fungus *(Ustilago may-dis)*, a very distant relative of regular mushrooms. The result could not be more unsightly: the kernels of the affected corn become discolored and distorted into bulging, irregular silvery-black masses that eventually ooze out a dark fluid (see Figure 9.4). The kindest thing I have heard said about them came from a friend of mine, Sarah Boardman, who told me that the misshapen kernels remind her of the Michelin man.

Few things look less appetizing than corn smut, yet in Mexico the affected corn is considered a delicacy. Several records indicate that Native Americans also used it and ranked it high in their cuisine. Corn smut is known in Mexico as *huitlacoche* or, in the words of certain food writers, the Mexican truffle. By themselves, the affected kernels have a strange taste that has been described as "metallic," "smoky sweet," or "a subtle blend of musky, earthy, mushroomy, and sweet corn tastes." Huitlacoche combines well with the basic ingredient of Mexican cuisine and can be used in a variety of ways. Coarsely chopped kernels serve as fillings in tacos, quesadillas, or in more elegant dishes such as crepes and puddings ("budines"). In the United States, this exotic-sounding food is available in gourmet Mexican food stores and, if farmers are savvy enough to take advantage of their misadventure, at farm stands during the corn season.

As would be expected, people in different parts of the world take advantage of the edible mushrooms that are peculiar to their region. Given the great variety of edibles that are found throughout the globe, the traveler may find local usage unfamiliar and surprising. For example, when Darwin traveled to Tierra del Fuego, he encountered the Yaganes, who remained scantily clad even in cold weather. The Yaganes subsisted on fish and shellfish, plus some berries and, as an important item in their diet, a mushroom that grows on the branches of the southern beech *(Nothofagus)* common in that area. This particular fungus, later called *Cyttaria darwinii,* is found only at the tip of South America, although some of its relatives grow in Australia and New Zealand. When mature, the specimens of this fungus look like yellow golf balls—they even have dimples. The Yaganes are long gone, and, according to the Argentine mycologist Irma Gamundi, so is the practice of using this particular species for food.

9.4
An ear of corn affected by corn smut.

Cyttaria darwinii and its allies are unusual among the mushrooms in that they contain large amounts of sugar. These mushrooms also carry yeast on their surface and eventually become fermented and emit a pleasant, beery aroma. Darwin commented that the Yaganes preferred the older and wizened specimens to the fresh young ones. They ate them raw. It occurred to me that the reason for this preference may have been that the mature specimens have a slight alcoholic content, a kind of fungal Baba au Rhum. I got the idea from reading that in neighboring Chile a related *Cyttaria,* called *llau-llau,* was used to make a kind of "chicha," a fermented, mildly alcoholic drink. I know of no other case where a fungus is not fermentor but fermentee. Llau-llau beer seems to have been abandoned in Chile, perhaps because of the ready availability of fine local wines. Nevertheless, llau-llau and another *Cyttaria* species, known locally as *digüeñe,* are commonly used for food in southern Chile and both are sold in markets to this day. They are eaten cooked or raw in salads. A Chilean friend of mine told me that as a boy growing up in the region of Valdivia, south of Santiago, he and his friends used cyttarias on trees as targets for slingshot practice.

In certain times and places, people made use of other parts of the fungal organism than the mushroom itself. In Chapter 8 I mentioned the *sclerotium,* a hard mass of tightly packed filaments. Sclerotia are survival structures: they store nutrients to help the fungus withstand harsh environmental conditions, such as low humidity. These structures are most often found underground, invisible to the casual picker, and are only detected when the soil under a mushroom is dug up. Morels develop from them, but in some cases sclerotia are sought for their own sake.

Various kinds of sclerotia have been used as food, notably by indigenous peoples in North America, Brazil, Nigeria, and Australia. People in northern Brazil used to search for buried sclerotia by probing with their feet for areas for ground that felt firmer, just as the truffles in the Middle East are hunted. One type of sclerotia was an important food source among Indians of the southeastern United States and was known as *tuckahoe.* Tuckahoes are brown, tuber-like structures with a scaly, bark-like covering, resembling in size and appearance an oblong coconut or a cassava. These sclerotia are white and translucent inside, have a firm texture, and, according to some writers, possess a "mushroom odor." They were eaten cooked and described as having a mild taste. What we know about this piece of American mushroom lore is somewhat fragmentary, in part because the term *tuckahoe* (and variants such as *tockawhoughe*) was used generically for several edible bulbous tubers, such as the wake robin, a type of *Trillium*. The connection between tuckahoe and fungi was suspected as early as 1762 by John Clayton, the author of one of the first books on American plants, *Flora Virginica*. The definitive identification of the species of fungus had to wait until 1922, when a North Carolina investigator, Frederick Wolf, found that a fungus now known as *Wolfiporia cocos* grew directly out of tuckahoes kept moist in the laboratory.

In Italy, a curious use was made of the sclerotia of a mushroom called *Polyporus tuberaster.* These large structures, the size of a grapefruit, cannot be eaten directly, because they are full of bits of wood, earth, and pebbles and have the consistency of stone. Rather, the *pietra fungaia* ("mushroom stone") used to be buried in a flower pot or other suitable container and watered twice a day. After about four days, edible mushrooms appeared and repeated crops could be obtained. By adding or withholding water, the gardener controls the growth of these "instant mushrooms." Sadly, extensive deforestation has put an end to this practice.

Mushroom production from the *pietra fungaia* has been known since Roman times and was discussed by classical authors, some of whom were properly mystified by it. Hermolaus, a botanist of fifteenth-century Venice, described the practice as a miracle: "Growing from a stone is a kind of mushroom miraculous in nature: if the stem is cut off at the base, another mushroom is born, as long as a part of the stem remains in the stone. Thus, the stone retains and increases its own fecundity. We believe that there is no other possibility to eat mushrooms that arise inside one's house."

What a pity that "mushroom stones" are not more readily available. Imagine being able to grow your own mushrooms indoors on demand. I once had the occasion to bring up the subject of sclerotia on a visit to the late Alexander Smith, an eminent mycologist at the herbarium of the University of Michigan. He presented me with an eight-pound, earth-encrusted specimen that resembled a cannonball. Its menacing looks took considerable explaining to the security guard at the Detroit airport. Alas, my attempts at inducing it to produce fruiting bodies failed.

Utsukushii ya ara utsukushii ya doku—kinoko

How beautiful,
Beautiful indeed,
The poisonous mushrooms!

Issa

KINGDOM OF VERSATILE PARTS

10

MUSHROOMS, THE MIND, AND THE BODY

"The ribs are yellow, just like the head," I was told on the phone by a lady whose eight-year-old daughter had been sick after eating wild mushrooms for lunch. She called me one October afternoon in order to get the mushroom identified at the request of the Boston Poison Center, for whom I was a consultant. How often I would get a "mushroom call" would tell me what kind of year we were having—I received few calls in a dry, mushroom-meager season. If it had rained a lot I was likely to get several calls a week, and they came at all hours of day and night. My job was not to tell people what they should do medically, just to try to establish if the mushrooms in question were likely to be poisonous. This is intrinsically difficult to do on the phone, and the job was not made easier by the understandable excitement of those calling.

Most of the people who would call me were worried about more than an upset stomach. Everyone knows that some species of mushroom are lethally poisonous or, quite literally, "mind-blowing." Fewer people are aware of the medicinal uses of mushrooms. Whether the effect is toxic, hallucinogenic, or therapeutic, the cause is the powerful chemicals found in some mushrooms. We tend to forget that the strongest chemical compounds are produced, quite naturally, not by organic chemists but by plants and fungi.

The chemical effects of fungi—be they welcome or feared—are duly noted in history. Ergot, a fungus that infects the rye plant and other cereals, has been blamed for much mischief. The sclerotium of this fungus forms small, dark, hard, banana-shaped bodies in the spikes of the cereal plant. People who eat ergot-contaminated rye are afflicted by a complex intoxication called "ergotism." This condition, which has afflicted millions throughout history, was known as St. Anthony's fire because of the burning sensation experienced by its victims (St. Anthony was believed to have power over flames and fire). There are two forms of ergotism, one that results in gangrene, with possible loss of limbs, and the other in convulsions and hallucinations.

In the past, changes in behavior in people who had eaten "ergotized" cereals have led others to suspect that the victims were under satanic spells. Attributing devilish powers to persons suffering from ergot poisoning may have been responsible for the Salem witch trials. This notion was proposed in 1976 in *Science* magazine (vol. 192, p. 21) by Linnda Caporael, who argued that the descriptions of the behavior of the accused also describe the behavior associated with ergotism. Furthermore, the weather in the summer of

1691 was damp and favorable for the growth of the fungus. This evidence seems circumstantial at best, but the proposal has led to a lively and unresolved debate over possible explanations for the accusation and punishment of witchcraft. It is clear, however, that ergotism has played a significant role in history. As recently as 1926, some 10,000 people in Russia got sick from such "bread poisoning." And for the sake of this fungus at least one battle was lost. The Russian cavalry was defeated by the Turks in the 1722 battle of Astrakhan because many soldiers had died or got sick from eating contaminated bread and their horses had gone "mad" from eating infected rye grains.

One of the challenges of consulting for the Boston Poison Center was getting an accurate description from nervous callers. Most people have no reason to be expert in describing the various parts of a mushroom, so in order to get any useful information my callers and I would fall back on common terms for size, shape, and color. I soon learned, for example, that "ribs" made more sense to the uninitiated than "gills" or "lamellae." Sometimes I was helped by unusual characteristics of certain species that could readily be told over the telephone. For instance, I got a fair number of calls regarding a hallucinogenic mushroom called "Laughing Jim" (*Gymnopilus spectabilis*)—which, as you might expect, induces laughter in some people. It is confused by the unwary with the edible honey mushroom, apparently because both are a light shade of brown and grow in clumps at the base of trees. It happens, however, that *G. spectabilis* has the unique property of turning green on cooking, an extraordinarily handy characteristic for identification via telephone. Here is a species that can be identified in a frying pan containing sautéed pieces drenched in oil and surrounded by bits of onion and garlic.

I remember one "spectabilis" call that got the better of me. I was at home and my wife took the call. She handed me the phone, looking mystified. The caller was an elderly lady who, it turned out, was on a "high." In the midst of giggles, she lost no time in coming to the point and wanted to know if she and her dinner guests were about to die. I tried to steer the conversation to my usual questions regarding the shape, size, and color of the specimens, and with some effort I did extract from her the information that she had served a rich pasta dish containing a few mushrooms she had picked off a tree that afternoon. Sure enough, the mushrooms had turned green on cooking. I couldn't get much more out of her because, amidst much tittering, she became quite impatient with me. When I suggested that she was probably not in real danger but should see a doctor anyway, she said: "You mean, we are not going to die [giggle, giggle]? Good-bye, then!"

Laughing Jim is bitter and highly unpalatable, requiring a determined experimenter to ingest it by itself. Accidental eaters, on the other hand, who have no knowledge or intention of "tripping" on this species, are likely to eat it only when the unpleasant taste is disguised by other ingredients.

As is true with all matters mycological, anything one says about Laughing Jim has exceptions. Specimens found in different regions vary greatly in their power; those in western North America, for example, are said to be uninteresting and to have practically no "kick." Even with active specimens, individuals who eat them vary in their reactions. Some people do not seem to be affected, whereas others have experiences that range from pleasant to scary.

Here is how one person described an intense encounter with Laughing Jim:

> My first thought when I closed my eyes was actually that I had opened them. I became aware of wide open spaces when I closed my eyes and various curving patterns in pinks . . . The patterns seemed to be in perpetual motion and to be alive, that is, I sensed they had a pulse of their own . . . My body began to feel numb or as if it were a body but not necessarily my own. This I experienced in a more poignant way when I turned over on my stomach in bed and caught a glimpse of my hand, which had some marks from the pillow case on it. It seemed to me the hand of a very aged person and I was startled. I felt some fear and so turned my hand with the palm facing me. It seemed so fleshy and white that I thought of a mushroom and immediately of mortality. I saw that I was made of flesh and that this was corruptible. Yet as I looked at my hand I saw it as a hand, not mine but one lent to me as it were to use in this life. It was meant to help me and I felt great pity on it because it was after all made of flesh and as

prone to decay as mushroom flesh. [A little later, the writer felt hungry.] I craved something raw and colorful so went to the crisper and took out two bright fuchsia-colored radishes and first set them on the table. How much they seemed like sperm or tadpoles, the way their central roots shot out like tails. They looked comical and I wanted to laugh . . .

On reading a draft of this chapter, my daughter, who knows that the drug culture has virtually passed me by, warned me that readers with "recreational" inclinations may be stimulated to go out and search for psychotropic mushrooms. I don't know whether or not this is likely to happen, but I feel obliged to warn those so inclined that consuming hallucinogenic mushrooms is both illegal and dangerous. Mind-altering mushrooms are considered controlled substances and are covered by antidrug statutes in most jurisdictions. Eating these mushrooms may be a risky undertaking for several reasons: psychotropic mushrooms may be confused with more poisonous ones, the amount of the active principle in a given specimen is unknown, mushrooms that are sold as hallucinogens are sometimes fortified with other drugs, and, at best, the effects are difficult to predict. The "trip" may well be to the emergency room!

In the sixties, the use of mushrooms as hallucinogens became second in popularity only to their consumption as food. Interest reached a peak with the introduction of the Mexican "magic mushrooms" to an eager North American public. The floodgates were unwittingly opened by a remarkable man, Gordon Wasson, who in the 1950s expanded a career as a banker in New York City to include the study of the role of mushrooms in human cultures. Much has been written by and about Wasson, who traveled extensively in pursuit of "ethnomycology," a term he coined. He followed up on an earlier finding by Harvard's ethnobotanist, Richard Schultes, who in 1939 identified some of the hallucinogenic mushrooms much used in Mexico as members of the genus *Panaeolus*. Schultes shed light on the difference between the hallucinogenic mushrooms and peyote, which is a cactus (*Lophophora williamsii*). Wasson became particularly interested in the Mexican mushrooms after he visited the Mazatec Indian *curandera*, Maria Sabina, whose *veladas* or seances he witnessed and taped.

Wasson championed the notion that hallucinogenic mushrooms were important factors in the early development of religion because they gave the shamans special powers conferred by psychotropic experiences. He and his collaborators claimed that not only Mexican religions but also the Greek Eleusinian rites, the vedic veneration of the deity Soma, and others were inspired by the consumption of hallucinogenic mushrooms. Although Wasson was unrelenting in pushing his interpretations to the limit and, in the eyes of some, overstated the influence of mushrooms, his research was exhaustive and extraordinarily well presented. Unfortunately, his two-volume *magnum opus, Mushrooms, Russia, and History,* written with his Russian wife Valentina, was published in a very expensive limited edition only and is not generally available to the non-specialist. It is an authoritative work that touches on nearly every imaginable aspect of the relationship between people and mushrooms. Later books by Wasson that were first published in deluxe editions were subsequently issued in affordable ones.

It cannot be said that Wasson proposed the most imaginative of connections between mushrooms and religion. That claim must go to the British biblical scholar John Allegro for his book *The Mushroom and the Cross*. Allegro maintained on linguistic grounds that Christianity was just an intricate ruse designed to disguise the real goings-on, which was the forbidden ritual use of mind-altering mushrooms. Allegro concluded that Jesus (and, for that matter, Abraham and Moses) never existed but were mythological figures contrived to conceal a mushroom-based mystery religion that had run afoul of the authorities. This notion has had few takers.

The general view that mind-altering substances were used early in human cultures for the purpose of divination and healing is based on many well-documented examples and has had wide acceptance. Compelling evidence for the use of mushrooms specifically is limited to only a few cultures, however. The most extensive documentation of ancient uses of mushrooms for sacred purposes comes from Mexico. The sixteenth-century codices, or chronicles, written by Spanish missionaries recording the practices of the Mexican Indians at the time of the conquest make clear mention of sacred mushrooms. They were known as *teonanacatl*, which means "dangerous mushroom" but which has been grandly translated as "flesh of the gods." The early Spanish missionaries considered eating teonanacatl to be equivalent to cannibalism, but their attempts to ban it were futile and they could not eradicate the use of "sacred mushrooms." These mushrooms cause a variety of hallucinatory and semi-conscious states, and the utterances of shamans under their influence were taken as warnings and prophecies.

Thanks to the studies of Schultes, Wasson, and others, we know that in Mexico these practices have continued uninterrupted to the present time. In one of the seances that Wasson taped in Mexico, Maria Sabina, the officiator, announced that a young man whose illness she was trying to cure was doomed. Indeed, he did die shortly thereafter. Schultes wrote that in recent times mushrooms were used not only by shamans for diagnostic and therapeutic purposes but also by professional diviners to locate stolen property, discover secrets, and give advice.

Hallucinogenic mushrooms, or "shrooms," belong to various taxonomic groups, but nearly all are agarics—mushrooms with cap, stems, and gills. Most of them have spores that are brown or nearly black in color. The best-known hallucinogenic mushrooms belong to the genus *Psilocybe*. Many species of *Psilocybe* and their allies usually grow on dung, but not all do. *Psilocybe* mushrooms contain psilocybin, a compound that is chemically changed in the body into the hallucinogenic substance psilocin (psilocybin contains phosphate, which must be removed for the molecule to enter into the cells of the nervous system). These mushrooms may be confused with deadly poisonous species, such as *Galerina autumnalis,* that grow on wood and have spores of a lighter, rust-brown color. "Shrooms" have invaded the information superhighway and there is a fair amount of electronic traffic concerning their desirability, if not their availability.

In any discussion of psychotropic mushrooms, the fly agaric, *Amanita muscaria,* deserves prominent mention. One of the most beautiful of all the mushrooms, *Amanita muscaria* has a bright, shining red cap with delicate white patches on the top. The fly agaric is clearly mushroom motif No. 1. It is everyone's idea of a mushroom and is depicted in children's books, nursery walls, or wherever a simple picture of a mushroom is needed. Perhaps it is so conspicuous because of the contrast of its white patches on a blazing scarlet background. This species has many regional varieties, and in eastern North America is usually a less eye-catching orange-yellow. Some kinds apparently have a greater mind-altering capacity than others. It should be noted that *Amanita muscaria* is not only hallucinogenic but also quite toxic, although, once again, these properties apparently vary with geography. My sample of people who have eaten the northeastern variety of *A. muscaria* is very small, three in number, but all of them reported that they became very scared and "felt like they had died." None of them had any intention of repeating the experience.

The best-documented use of *A. muscaria* as an intoxicant comes from the Kamchatka peninsula in northeastern Siberia, where the Koryak tribesmen and other nomadic herders have been known to western visitors to partake of it since the seventeenth century. Ingestion of this mushroom is reported to cause animation, exhilaration, and visual hallucinations. Eventually, a deep slumber ensues, followed by headache and nausea. As with the Mexican shamans, the altered states of mind supposedly allowed the Koryak shamans to communicate with the supernatural, divine the future, and diagnose illness. However, the fly agaric was also used recreationally by the rest of the people, especially on festive occasions. Since the 1700s, travelers to the region have reported a distinctive way of consuming this mushroom: in order to make the most use of it, the Koryak drank the urine of those who had eaten it. The introduction of vodka resulted in the diminished usage of this mushroom in Siberia, which seems hardly surprising. According to the narration of some travelers, the mushrooms tasted so bad that they had to be rolled into small sausage shapes that could be popped down the throat. It is not entirely clear if the urine was drunk just to recycle the active principles or also to remove the substances responsible for undesirable side effects, or both. The active substances in the fly agaric have been identified as muscimol and ibotenic acid. The more important of the two, muscimol, is in fact excreted unaltered by the kidney. The fly agaric contains only small amounts of the alkaloid muscarine, which got its name from this mushroom but is not responsible for its pharmacological properties.

In Siberia, the distinction between the sacred uses of "magic" mushrooms and eating them for enjoyment was often blurry. The same can be said for the Kuma of New Guinea, whose consumption of hallucinogenic mushrooms was not limited to just a few members of the tribe. Nor were the mushrooms prepared by a special ritual. The unusual thing about the Kuma experience is that ingestion of the mushrooms led to a different response in men and women. Collectively called "nonda," the mushrooms have not all been definitively identified, and several species are consumed. After eating nonda, the men got very excited and combative, to the point of threateningly brandishing their weapons and even lightly wounding others. The women, on the other hand, became flirtatious and provocative, boasting of sexual exploits with members of their husbands' clans. Under the influence of the mushrooms, married women were decorated with their best feathers and were allowed to perform dances reserved for men and unmarried women. Apparently, the women under the influence of the mushrooms were not considered responsible for their actions. In general, eating psychedelic mushrooms does not necessarily lead to heightened sexuality, although, if it did, the effect would be attributed to the release of inhibitions. A connection between sexuality and mushrooms was reported from North America, in 1626. In a letter to his brother in France, the Jesuit Pére Charles l'Allemant wrote that the Algonquins believed that "after death they go to heaven where they eat mushrooms and have intercourse with each other."

The fly agaric as an intoxicant has been well known since antiquity in various parts of Europe. To this day, people in Hungary call it *bolond gomba,* the "mad mushroom," and similar terms are used in Austria. There may even be a deeper connection between the term *fly agaric* and its hallucinogenic power. Flies that feed on pieces of this mushroom become drowsy and are easily killed, and it's true that the name is probably derived from this fact. The species is linked to flies in various languages: it is called *tue mouche,* or "fly-killer," in French; *mukhomor,* or "fly-killer," in Russian; *Fliegenpilz,* or "fly-mushroom," in German; and *aka-haetori,* or "red fly-catcher," in Japanese. It is also the case that flies and other insects have long been associated in people's minds with intoxications and madness. The Russians call one who is drunk *pod mukhoj,* "being under fly." Beelzebub, later called Satan, was also known as "Lord of the Flies." All over Europe, a large number of legends were told concerning the fly agaric. In the New World, on the other hand, the use of this particular mushroom seems to have been confined to the Ojibway of the Great Lakes regions, who incorporated it in ancient ceremonies.

One of the hallucinatory effects of *A. muscaria* is to distort the size of objects and their surroundings, making them appear larger or smaller. Visual hallucinations of this sort may explain the episode in Lewis Carroll's *Alice in Wonderland* in which Alice shrinks or elongates depending on which side of the mushroom she nibbles. The magic mushroom is the one on which the caterpillar sits ostentatiously while smoking a water pipe. There is evidence that Lewis Carroll was well informed about the hallucinogenic power of mushrooms and was aware of the visual distortions caused by the fly agaric. In an earlier version of *Alice,* the two different effects of the mushroom were attributed to eating separately from the stem and the cap, an intriguing notion that was dropped from later editions.

I can report an exciting, if unwitting, experience with *Amanita muscaria.* On a Sunday morning, a taxi driver delivered a mushroom specimen to my home at 6:30 A.M. It had been sent to me from the emergency room of Boston's Massachusetts General Hospital. The taxi driver was so convinced that this was a matter of life or death that when my sleepy son opened the door, he dashed into the house and loudly asked him where I was. He ran up the stairs and walked straight into my bedroom, bag in hand! After he left and I recovered my wits, I called the hospital to piece the story together. It turned out that the day before a young man had indulged in so many different drugs that he felt the need for medical help. He had taken uppers, downers, smoked marihuana, drunk alcohol—in other words, the works. He felt sick but feared that a hospital would not take him seriously, so he decided to present himself to the emergency room clutching an *A. muscaria* he had saved in his refrigerator for a future occasion. He proudly presented the specimen and announced that it was what he had consumed. Actually it was about the only thing he hadn't!

The seemingly casual use of hallucinogenic mushrooms

causes me discomfort. I am not well grounded in this subject, but it seems easy to accept that in the context of
close-knit societies, mind-altering experiences have contributed positively to human development. The spiritual, psychological, and esthetic components of such experiences
have been aptly celebrated by scholars of the subject. The
deeper meaning of "magic mushrooms," however, is likely
to be lost without the benefit of long and well-established
traditions. I would think that the proper "set and setting"
are difficult to re-create. Personally, I don't want to see mushrooms lumped together more or less indiscriminately with
all sorts of white powders and pills. They deserve better, and
so do we.

Fungi have body-altering as well as mind-altering properties. The ergot that causes hallucinations and other psychological effects has serious physical effects too. It contains
ergonovine, which is used to constrict blood vessels after
childbirth or abortions, and ergotamine, which is helpful
in the prevention and treatment of migraine. (In addition,
ergot contains a less potent relative of LSD, lysergic acid,
but I am discussing the *medicinal* uses of fungal pharmaceu-
ticals now.)

The list of medicinal compounds derived from fungi is
long. It includes most of the antibiotics, the immunosup-
pressants used in organ transplants, and cancer chemothera-
peutics as well as the alkaloids of ergot. Except for ergot,
most of the products in current use are made by filamentous

molds (which are funguses but do not produce mushrooms). Plus, there are many compounds derived from mushrooms for which therapeutic claims have been made and which await further investigation. These studies are the subject of a huge literature. From China come studies that suggest that compounds contained in wood ears (*Auricularia polytricha*) are active in thinning blood. Japanese, Chinese, and Korean researchers have claimed that lentinan, a substance contained in shiitake (*Lentinus edodes*), is helpful in the treatment of cancer, heart disease, high blood pressure, and diabetes. Lentinan and other polysaccharides from mushrooms have been shown to stimulate the immune system, but it is not known if they are any more useful than other substances with similar activity. The last word on the subject has not been spoken.

Many examples of the use of fungi in medicine come to us from antiquity. By far the most ancient written records are those from China. The oldest book on medicinal substances was written in the first century B.C. and is known as *Shen Nung's Herbal*. It contains detailed descriptions of the beneficial effects of a number of fungi and was followed by considerable writing on the subject. In the West, fungal products were used in various ways; some were taken internally and others applied to the skin. Since Hippocrates's times and into the nineteenth century, slivers of bracket fungi were burned over afflicted areas of the body to cause blistering. For a long time inducing blisters was believed to alleviate pain and fight inflammations. Along with bleeding, blistering was one of the most widely used medical procedures in the West until well into the twentieth century.

Among the medicinal uses of mushrooms, Pliny the Elder mentions that boletes cure "fluxes of the bowels," remove freckles, help with sore eyes, and alleviate dog bites. A bracket fungus called "agarick" or "Agaricum" (*Polyporus officinalis*), taken internally, was reputed by the second-century A.D. Greek military surgeon Dioscorides to be the "panacea" or "universal medicine," good for curing fevers, colic, jaundice, asthma, sores, kidney disease, hysteria, and fractured limbs, among many other things. Dioscorides mentions that it will also cure snake bites. Alas, the only proven fact regarding the therapeutic powers of *P. officinalis* that survives is that it is a strong purgative. The repute of fungi in medicine lasted well into the Renaissance because of the continued reliance of physicians on the writings of Greek and Roman medical authorities. The famous herbal written by Gerard in 1597 mentions among "agarick's" curative powers that: "It provoketh urine and bringeth down the menses; it maketh the body well coloured, driveth forth wormes, cureth agues, especially quotidians and wandering fevers, and others that are of long continuance, if it be mixed with fit things that serve for the disease: and these things it performes by drawing forth and purging away grosse, cold and flegmaticke humours, which cause the diseases."

Some of these ancient uses have survived for a long time and have continued well into this century, in parts of Scandinavia, for example. Occasionally, medicinal uses for fungi and their products surface anew. Recently Kombucha, a fermented fungal concoction of Manchurian origin, has taken "agarick's" place. According to newspaper reports, Kombucha's adherents maintain that it helps with cancers, lowers the blood pressure, eases the pains of arthritis, and restores the original color to graying hairs. It has been used by AIDS patients. Kombucha is made by placing a disk-shaped blob consisting of a mixture of yeasts and bacteria on sweetened tea and letting fermentation take its course. In time, the microorganisms will make another disk (a "baby"), which can be propagated many times. The fermented drink tastes somewhat like hard apple cider and is high in acidity. Kombucha is known by many other terms, including "Manchurian mushroom," which is an unusually loose extension of the term "mushroom."

Native Americans had extensive knowledge of fungi and had dozens of applications for them. For example, puffballs were used to stanch bleeding in wounds or as poultice for sores. In general, these uses appear to be more effective than those advocated by the European physicians. Both in North America and in Europe it was appreciated that the fumes given off by burning puffballs had an anesthetic effect, but this property seems to have been used mainly for smoking out bees from beehives. In many North American tribes, dry puffballs were also squeezed into the nostrils to stop nosebleeds, but this turns out to be a bad idea because inhalation of the fine spore powder can cause inflammation of the lungs. The Rappahannock Indians of Virginia were aware of this danger and called the puffball powder the "devil's snuff." More recently, a group of fifteen teenagers in southeastern Wisconsin in April 1994 inhaled puffball spores in the belief that doing so would give them a high. Seven of them came down with a severe form of pneumonia that required hospitalization. It took a lung biopsy to establish this unusual diagnosis. One of the teens had part of a lung removed. There is some basis for the notion that puffballs may be hallucinogenic, because some relatively unusual species indeed are, but I don't know whether these young people were aware of this information or just thought that any wild mushroom may give them a "trip."

In addition to being used to induce physical and psychological changes, mushrooms have been appreciated for their alleged aphrodisiac powers. It is a widespread belief in many cultures that certain foods are sexually arousing. Among the mushrooms, the most obvious candidates are the phallic-looking stinkhorns (see Chapter 9), in which case a connection with human sexuality requires no imagination. Perhaps some mushrooms do have aphrodisiacal effects: a Japanese investigator, A. Kaneda of Tohoku University, reported that shiitake-fed female rats grew larger and pinker nipples.

In at least one case, the desired effect seems to have been indirect. The eighteenth-century Swedish founder of modern taxonomy, Linnaeus, described a special use for a bracket fungus (*Trametes odora*) among the Laplanders. He wrote: "The Lapland youth carefully keeps it in a pouch hanging in front of his pubes (*in marsupio ante pubem pendulo*), that its lovely odor may render him more acceptable to his fa-vored maiden. O whimsical Venus! In other regions you may be treated with coffee and chocolate, . . . jewels and pearls . . . music and theatrical exhibitions; here you are satisfied with a little withered fungus!"

11

MURDER AND MORE
MUSHROOM MAYHEM

Second only to "Is this one good to eat?" the question I am most often asked about mushroms is: "How can you tell a poisonous from an edible one?" The answer is: "With difficulty." There are no reliable rules. No generalization applies in every case, least of all the old saws that poisonous mushrooms make silver spoons turn black or that a mushroom that can be peeled is edible. The most poisonous of all mushrooms, the deadly amanitas, fail both tests entirely: their caps can be peeled, and they do not turn silver black. What makes mushrooms insidious is that many of the poisonous kinds resemble edible ones. A little bit of knowledge is a dangerous thing: anyone intending to eat a mushroom must know it is safe *with certainty*. Just being "pretty sure" is recklessness. I must emphasize that this book is not intended as a primary source for the identification of edible mushrooms. If you are interested in eating wild mushrooms, you must go elsewhere for information and advice on identification. (I have listed some suggestions at the end of the book, under "Resources.") A very good place to start is your local mushroom club, the members of whom are apt to be fastidious about proper identification. Consider this warning, by one Hollis Webster, published in the bulletin of the Boston Mycological Club in 1897: "Even among the members of the Mycological Club—who ought to know better—there have been one or two instances where sickness has followed an ill-judged meal; and more than one member—fortunately so far without result—has carelessly confounded a kind to be avoided with one recommended. Such mistakes ought to work a suspension of the privileges of membership; at least any one careless about amanitas should hardly expect to remain an active member."

A dozen or so people in the United States are severely poisoned from eating wild mushrooms every year. A few of them die. Although the exact number is not known—the Centers for Disease Control do not keep records of mushroom poisoning—it is less than the number of deaths due to bee stings or lightning. These cases represent a minority of all the mushroom poisonings; most poisonings are not nearly that severe. In most instances, the person who eats a poisonous mushroom suffers an unpleasant but relatively mild and short-lived intestinal disorder, much like the symptoms caused by food poisoning. Nevertheless, each poisoning is a surprise, for there is a strong taboo in this country against eating mushrooms not bought in a store. Nearly everyone seems to know that wild mushrooms can be dangerous and should be avoided at all costs.

Some of my friends think I am playing with fire when I indulge in a plateful of wild mushrooms. I tell them that I always obey the rules of only eating mushrooms I know for sure are safe and only sampling one or two bites when trying out a new species. However, I recognize that it may be tempting to break these rules. Imagine finding a resplendent twenty-pound specimen of the chicken mushroom (*Laetiporus sulphureus*) and proudly bringing it home. You've eaten this species of mushroom plenty of times, so has your family, and you believe it is as safe as mother's milk. That evening you have a party at your house, and you sauté a whole mess of the mushroom and pass it around to your guests as a delectable and novel finger food. You are sharing the delights of your hobby with your friends and want to prove to them that it is safe and enjoyable. Sure enough, your skills as a gatherer are much admired. Some people just have a small bite but others like the mushroom so much that they eat a lot of it. Probably nothing will happen, but what if some of your guests are allergic or otherwise sensitive to this mushroom?

Mushroom poisoning is hardly new. It is mentioned in the earliest writings on mycology by the ancient Greeks and Romans. To quote Pliny the Elder: "Among those foods that are eaten carelessly, I would place mushrooms. Although mushrooms taste wonderful, they have fallen in disrepute because of a shocking murder. They were the means by which the emperor Tiberius Claudius was poisoned by his wife Agrippina. Thus, she gave the world a poison worse still—her own son Nero."

In defense of mushrooms, some historians proclaim their innocence as the agent of death but concede that they could be used as a handy vehicle for some other poison. Pliny himself states that mushrooms were an accommodating instrument for other poisons (*venenis accommodatissimi*), which for him was reason enough to avoid them altogether. Perhaps that is why mushrooms were the only food that was prepared by the host at Roman banquet tables. This practice did not ensure safety, but at least it might cast suspicion on a possible perpetrator.

Mithridates VI (also known as the Great), king of Pontus in Asia Minor in the first century B.C., tried to become resistant to poisons by taking them in gradually increasing doses. It is not known if mushroom poisons were included in the menu, but the method may well have worked. In 1976, the Swiss biochemist G. Floersheim discovered that injecting mice with small amounts of an extract of deadly mushroom (*Amanita phalloides*) protected the animals from a later injection of a usually lethal dose. The reason for this acquired drug tolerance has not been definitively established, but the immune system is not responsible for it. No one would suggest that someone who enjoys wild mushrooms should emulate Mithridates and eat increasing "doses" of poisonous varieties—unless, perhaps, they have as many enemies as he did.

Mushroom poisonings make good stories, and several mystery writers have used them in murder plots. Perhaps the best known of these stories is *The Documents in the Case,* by Dorothy L. Sayers and Robert Eustace. In this book the victim succumbs to muscarine, a toxic compound ingested, it would appear, along with the fly agaric, *Amanita muscaria.* The murderer is caught because an insightful chemist found out that the victim was killed by a synthetic form of the poison, which could only have come from a malevolent, not a natural, source. The facts of the story are not quite right: despite its name, this mushroom does not contain much muscarine, and besides, muscarine is not that toxic. On the other side of the coin, some writers have used mushrooms to prove that crime *does* pay. In a film by Sacha Guitry, *Le Tricheur* ("The Trickster"), the protagonist as a small child is sent to bed without his dinner because of his transgressions. The rest of the family sits down to a meal and, sure enough, all die of mushroom poisoning. The survivor learns from this lesson and is inspired to go on to a successful life of crime.

As well as serving as weapons of a sort, mushrooms may also help in the detection of corpses. The Japanese mycologist Naohiko Sagara reported in the British journal *Nature* (August 26, 1976) that he found many specimens of a mushrooms called *Hebeloma vinosophyllum* growing in a small area above the buried carcasses of a dog and a cat. This species seems to "fruit" best when provided with nitrogen-containing compounds arising from the decomposition of animal bodies. The author speculates that "a victim of homicide buried in soil could be located using the fruiting bodies of this or some other fungus, assuming the burial was shallow." I do not want to give the impression that this mushroom killed the animals, just that it makes good use of the nutrients emanating from corpses. A similar species of mushroom (*H. syriense*) is found in North America and Europe, where it's known as the "corpse finder." Perhaps the day will come when police detectives will be encouraged to study mycology.

Other dangers may accompany a meal of wild mushrooms. In the aftermath of the Chernobyl disaster, mushrooms in Russia and other countries affected by the fallout were found to accumulate dangerous levels of radioactivity. Mushrooms can absorb from the substrate large amounts of heavy metals, which in this case included dangerously radioactive isotopes. This ability of mushrooms to concentrate metals is one reason to avoid eating those gathered from the verges of highways, especially in countries where leaded gasoline is sold. Other environmental hazards associated with mushroom hunting have nothing to do with poisoning. A news item from Associated Press (October 20, 1994) reported that a 50-year-old man collecting wild mushrooms in the hills north of Madrid was fatally gored by a fighting bull that had escaped from its pen.

"If it tastes bitter, I don't eat it, otherwise I do." I have heard this from people of all ages and origins who ought to know better. Some credit their grandparents (who "knew everything about mushrooms"), others believe in making their own rules. I have tried to argue against these notions only to be told: "I have been picking mushrooms for the last forty years and I have never had any trouble." "You mean, not even a stomachache?" "Oh, yes, I've had lots of those!" As I mentioned earlier, immigrants from Laos and other Southeast Asian countries have not been not so fortunate. A number of these newcomers have suffered from life-threatening poisonings after eating wild mushrooms in California and Oregon. According to the New York mycologist Gary Lincoff, one of the deadliest mushrooms in North America, the destroying angel *(Amanita virosa),* looks very much like an edible mushroom that is sold in local markets of their native lands *(Amanita princeps).* Several of the victims had to undergo liver transplants.

Most of the people who are poisoned from eating wild mushrooms fall into one of three categories:

- Toddlers, grazing on lawns.
- Recreational drug users, who think that any trip is a trip. Among the favorite mushrooms of this group is *Amanita muscaria,* consumed despite warnings about its toxicity.
- People who have come to the United States from countries where mushrooming is popular. There are

no statistics to confirm this, but I have the impression that immigrants from the Baltic nations, Italy, and Southeast Asia are especially susceptible to the danger of encountering toxic North American mushrooms resembling species that are safe to eat in their home regions.

With the exception of morelling traditions in the Midwest and the West, there is little wild mushroom lore in North America. Native Americans consumed a variety of mushrooms, but not much information about the different species seems to have been passed on to the European settlers. I had high expectations of broadening my knowledge of New England's mushrooms when I received a call concerning an elderly Yankee gentleman living in rural New Hampshire. He had been sent to the Massachusetts General Hospital because he had symptoms of mushroom intoxication. Fortunately, he recovered quite rapidly and I had the chance to ask him questions. I was sure that at long last I had found a well of folk wisdom passed down by word of mouth for generations. Alas, when I asked him how long he had been picking wild mushrooms—"For the last four years"—my hopes were dashed.

One of the biggest mistakes people make about the danger of toxic species is that deadly mushrooms work very fast, as though after one bite a victim would just drop dead. The very opposite is true: the truly dangerous mushrooms are slow-acting and typically do not cause symptoms until six to twelve hours or longer after ingestion. Species that cause milder intoxications make people sick in two hours or less. The faster the onset of symptoms, the less severe is the outcome. Facts about mushroom poisoning are not easy to come by, as several factors make mushroom toxicology a difficult science. Most cases occur in isolation, although picnics and other communal feedings have at times affected several people at once. Often, the mushroom is poorly identified, or more than one species may have been eaten at the same time, or the diagnosis did not distinguish between the mushrooms and spoiled food as the cause of the poisoning. In defense of the mushrooms, the Reverend M. J. Berkeley, an eminent nineteenth-century British mycologist, wrote that "it is not always the poisonous properties of species that are to be questioned. A man after a long day's fast, for instance, eats a pound or two of mushrooms badly cooked, and frequently without a proper quantity of bread to secure their mastication, and then is surprised that he has a frightful fit of indigestion."

Mushroom poisonings fall into several categories, ranging from the life-threatening to the merely annoying. The most dangerous sort, the one that causes most fatalities, is the ingestion of toxins known as amanitins. As the name suggests, the toxins are found in members of the *Amanita*, a group of stately mushrooms that are relatively easy to identify, at least to the level of genus. In addition to caps, stems, and gills, the amanitas have a ring around the stem and, at least when young, they are decorated with white or colored patches on the top of the cap. In addition, the base of the stem of an amanita is either encased in a cup or has remnants of the tissue derived from a cup. The name "death cup" is appropriately used for one of the most deadly of the amanitas, *Amanita phalloides*. Amanitas are particularly beautiful and cause much fascination. The most common of the deadly amanitas in eastern North America, the pure white destroying angel *(A. virosa)*, is so stately and resplendent that a Wisconsin mushroomer, Jessie Saiki, called it "death masquerading as a virgin bride." Another *Amanita* that is white when young, *A. ocreata*, is the "destroying angel" of the West Coast.

Not all amanitas are deadly, in fact some are edible and delectable. In ancient Rome, a luminous orange amanita, Caesar's mushroom *(A. caesarea)*, was particularly prized. On the West Coast, *A. calyptrata* is avidly sought by Italian-Americans, who call it "coccora." Edible and poisonous amanitas are not that difficult to tell apart, but there is always an element of chance and, as always, you should never eat a mushroom you are not certain is edible. No one should eat amanitas without a "license," by which I mean considerable knowledge and experience in mushroom identification. One should even choose one's imprimatur with caution. The 1991 edition of the venerable French *Petit Larousse Encyclopédie* had to be recalled because the illustrations of deadly amanitas were labeled with the symbol for "indifferent" rather than "poisonous." No less than 180,000 volumes had been distributed, some 250 students were hired to visit the 6,000 stores in France, Belgium, Switzerland, and Canada where the book was being sold, in order to place sticky labels with the correct wording on the right page. A notice was then placed on each cover denoting the altered volume as a "new edition"!

Eating a single specimen of a poisonous *Amanita* may be enough to cause serious trouble—even death. The symptoms start after six to twelve hours, with violent nausea, diarrhea, and cramps that often abate after a day. Because there is such a long lag between eating the mushrooms and the appearance of symptoms, washing out the stomach does not alleviate the symptoms. Signs of extensive damage to the liver and other organs appear around the second or third day. The mortality rate among victims used to be around 30 percent of those who experienced serious symptoms. Many treatments have been tried in the past and failed. Nowadays, a better understanding of the physiology of the intoxication has allowed physicians to cut the mortality rate to 5 percent or so. Treatment involves restoring the electrolyte and sugar balance in the patient and, according to one protocol, removing as much of the toxins as possible by passing a tube into the patient's duodenum and pumping out the stomach bile. Because kidney function is impaired, toxins, instead of being metabolized through the urinary system, recirculate through the liver, are excreted through the bile, and can then be reabsorbed in the intestine. That is why the stomach bile is sometimes removed. In more advanced cases, physicians have had to resort to liver transplants. No matter how much

11.1
α-Amanitin, the toxin responsible for most serious mushroom poisonings. It is found not only in certain amanitas but in some other poisonous mushrooms as well.

$$
\begin{array}{l}
\text{OH} \\
\text{H}_3\text{C} \quad \text{CH—CH}_2\text{OH} \\
\text{CH} \\
\text{HN—CH—CO—NH—CH—CO—NH—CH}_2\text{—CO} \\
\text{OC} \qquad \text{H}_2\text{C} \qquad\qquad \text{NH} \quad \text{CH}_3 \\
\text{H} \quad \text{CH} \qquad\qquad\qquad \text{HC—C} \\
\text{HO} \quad \text{N} \qquad \text{O=S} \quad \text{N} \quad \text{OH} \quad \text{CO} \quad \text{CH}_2\text{—CH}_3 \\
\qquad\qquad\qquad \text{H} \\
\qquad\qquad \text{CH}_2 \\
\text{OC—CH—NH—CO—CH—NH—CO—CH}_2\text{—NH} \\
\text{H}_2\text{C—CO—CONH}_2
\end{array}
$$

we have learned about their treatment, amanitin intoxications remain a medical emergency.

The amanitins are strange compounds. They consist of about eight amino acids in the shape of a figure eight, a molecular pretzel (Figure 11.1). They work by inhibiting a key enzyme in the nuclei of all cells, one of the RNA polymerases. As a consequence, messenger RNA is not made, proteins are no longer synthesized, and the cell machinery goes awry. The reason the liver is especially sensitive is that, since it is one of the most metabolically active organs of the body, it carries out a lot of protein synthesis and becomes a target for drugs that inhibit this process.

The question arises, why do poisonous mushrooms make toxins? The simple answer—they need them to keep from being eaten by animals or humans—is not very satisfying. It's true that the poisonous mushrooms seem to be avoided by squirrels, chipmunks, and deer, but if this were so important one would expect poisonous mushrooms to be more abundant than the innocuous ones and eventually to become the dominant species, which is clearly not the case. In fact, it is possible that being eaten helps disperse the spores in the environment, as spores are not easily digested and may pass through the intestinal tract unharmed. The honest answer is that we don't know why mushrooms make toxins, but neither can we explain a large number of compounds manufactured by living things, such as many of the alkaloids and pigments made by plants. Why do coffee plants make caffeine, and why are roses red?

A clearer question is: how do poisonous mushrooms manage to resist the action of the toxins they make? The mushrooms that synthesize amanitins have enzymes that are similar to our own enzymes. However, it has been shown that the RNA polymerases of these and other mushrooms are more resistant to the toxins than the same enzymes of animals. Another interesting point is that the demarcation between mushrooms that produce amanitin those that do not may not be as sharp as is usually believed. In a tantalizing 1982 study, T. Romeo and J. F. Preston of the University of Florida concluded that amanitin production may be more widespread than is commonly recognized because they found that some non-amanitin-containing mushrooms can degrade these toxins. Thus, some species may in fact make significant amounts of amanitins but may break them down as well. The next question that arises is, why do these mushrooms bother to make amanitins if they are then going to get rid of them?

Amanitins are certainly not produced only by the members of the genus *Amanita*. Several other mushrooms contain significant amounts of these toxins and have been responsible for serious intoxications. One of the genera implicated is *Galerina*, fairly insignificant-looking brown mushrooms that usually grow in clumps on fallen logs. People become poisoned because galerinas are sometimes mistaken for the honey mushroom (*Armillaria mellea*) or the hallucinogenic *Psilocybe*, with which they share a few features only. Experienced mushroomers would normally not mistake these two kinds, but if a beginner focuses only on a few characteristics—such as the fact that these kinds all have a ring on their stem, are brownish in color, and tend to grow in clumps—it is not unlikely that the specimen could be misidentified.

Close relatives of edible species may contain large amounts of amanitins. Once again, I must emphasize the need for caution. A knowledgeable mushroomer in upstate New York died some time ago from eating a sandwich made with a relative of the parasol mushroom (*Lepiota procera*). It turned out that he identified the specimen correctly but he didn't know for sure if it was safe to eat. He assumed that this small, infrequently found species, *Lepiota josserandii*, was edible. It turns out instead that this species contains deadly amounts of amanitins. The victim suffered his tragic fate because at the time there was practically no information in this country to contradict the belief that most lepiotas are edible. The most widely known poisonous member of this group is a species found very often on lawns. It is *Chlorophyllum molybdites,* which causes gastrointestinal upsets. This species may be confused with the edible meadow mushroom *(Agaricus campestris),* but it can be distinguished from *A. campestris* and practically all other mushrooms by its green spores.

More common than life-threatening poisonings are those that cause gastrointestinal disorders resembling the familiar types of food poisoning. Symptoms show up half an hour to four hours after eating the mushrooms. There may be vomiting, nausea, and cramps, all of which may be intense and accompanied by diarrhea. The symptoms are not, generally speaking, as severe as those that follow amanitin poisonings early on. The list of mushrooms that can cause these symptoms is very long, and it includes varieties in most of the major groups (see the accompanying table). There seems to be considerable variation within a mushroom species regarding the reaction it elicits in people. Some individuals are sensitive to mushrooms that are widely considered edible. In fact, the list of wild mushrooms that have been reported to cause symptoms of intoxication in some person includes all the species considered to be highly edible. What's going on? We don't really know. The trouble is that "mushroom poisonings" are hard to document and the information we have is based largely on anecdotes. There are practically no data to suggest what kind of individual reactions these are. The word *allergy* is bandied about, but without much evidence.

Mushroom Poisoning

Symptoms	Toxins	Species
Vomiting, nausea, diarrhea early on. Later, bloody diarrhea, abdominal cramps, liver and kidney failure. May be fatal.	Amanitins (cyclopeptides)	*Amanita virosa; A. phalloides;* other amanitas; some lepiotas, galerinas
Symptoms similiar to above. May be fatal, especially for small children.	Methyl hydrazine	False morel (*Gyromitra esculenta*); related ascomycetes
Confusion, euphoria, loss of coordination, sweating, hallucinations, drowsiness.	Muscimol, ibotenic acid	Fly agaric (*Amanita muscaria*); *A. pantherina;* others
Perspiration, tears, salivation.	Muscarine	*Clitocybe dealbata;* some inocybes; some boletes; others
Nausea, vomiting, diarrhea, usually lasting 1–2 days.	Not well known	*Agaricus hondensis; Chlorophyllum molybdites; Boletus satanas;* entolomas; hebelomas; many others
Visual hallucinations, sometimes anxiety, other psychological changes.	Psilocybin, psilocin	*Psilocybe caerulescens;* other psilocybes; *Panaeolus* species

The most wholesome of legumes, grains, fruits, or meats may cause idiosyncratic reactions in some people. The good news is that such events are relatively rare.

I remember a particular instance of a gastrointestinal disorder from my days as a consultant to the Boston Poison Center. A Rhode Island man had eaten a mushroom for dinner and came down with a touch of stomachache. He contacted the Providence Poison Center, where it was decided that what was left of the specimen had to be sent to Boston for identification. Not so fast! The Rhode Island state trooper dispatched on this occasion could only deliver the mushroom to the border, where his Massachusetts counterpart had to pick it up and bring it to me. We seem to be able to extradite wild mushrooms as long as they are conveyed through official channels. It turned out that the culprit belonged to a species called *Tricholomopsis platyphylla,* whose ability to cause gastrointestinal irritation had been first reported by Dr. Roger Goos at the nearby University of Rhode Island.

The list of different kinds of mushroom poisonings is quite extensive. One species of false morel (*Gyromitra esculenta*) contains a compound that turns into methylhydrazine, a highly volatile toxin that has been used as a rocket fuel. Cooked, the mushrooms are safe, but they may be exceptionally hazardous to those that inhale the fumes during their preparation. Thus, the cook may get sick, whereas those eating the cooked mushroom will escape with impunity. Gyromitras should, of course, never be eaten raw, and

some cautious experts believe they should be avoided altogether. Some of the inky caps (*Coprinus atramentarius*) contain an unusual amino acid (coprine) that inhibits the enzyme that metabolizes alcohol. I mentioned already that this
compound mimics the action of the drug disulfram (popu-
larly known as Antabuse). People who eat the mushroom
and ingest alcohol may suffer from rapid heart beat, redden-
ing of the skin, and other unpleasant symptoms.

Throughout history, many people have attempted to de-
toxify poisonous mushrooms and make them fit to eat. Un-
doubtedly there are several ways to do this, such as cooking
or drying gyromitras, but some of these procedures are dangerous and, in some cases, strain one's sensibilities. A French
experimenter, for instance, concocted a potion that included
as one of its ingredients chopped raw rabbit stomachs. An
interesting report on rendering poisonous mushrooms edible focuses on a little-researched topic, the use of mushrooms by African-Americans from the time of their arrival in America onward. An 1898 publication of the U.S.
Department of Agriculture (by someone named Coville) tells
that African-American women in Washington, D.C., were
divided regarding their opinion of the edibility of the fly
agaric, *Amanita muscaria*. Most avoided it, but some would
eat the species if it was prepared in a complex recipe that
included removing the gills, peeling the surface of the cap,
boiling for a long time in salt water, and steeping it in vinegar. Although the procedure apparently worked, the author
adds a word of discouragement to anyone inclined to eat a

poisonous mushroom in any form, "particularly at a season when excellent non-poisonous species may be had in abundance."

My personal experience with mushroom poisoning is based largely on the calls referred to me through the Boston Poison Center. I have learned a great deal from these calls, and sometimes I have done even better than that. Once I was called regarding the edibility of some funny-looking mushrooms growing under an apple tree. Since I firmly stick to the rule never to attempt to identify specimens for the table over the telephone, I invited the caller to bring me her specimens for inspection. She came with two pounds of morels, half of which became my identification fee. Call anytime, Ma'am!

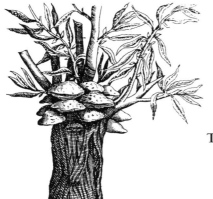

THE TRAIN WRECKER AND OTHER
STURDY MUSHROOMS

Many a devoted mushroomer makes forays for books as well as mushrooms. Among my usual haunts when I lived in New England was a particular antiquarian bookstore. The store is now long gone, the building in downtown Boston that housed it on the first floor destroyed by the wrecker's ball. I used to visit it on my lunch break, as I made my rounds of second-hand bookstores. Most of the other stores I frequented had bookcases all the way up to the ceiling, some made of regular shelves, others just stacks of simple crates. This particular store was different: it was wide open, with a few low bookcases on the sides and a couple of large tables in the center, covered with pamphlets lying face-up. Sounds like a modern design, but the description is misleading: everything in the store was of venerable age and the contents overlaid with a coat of ancient dust.

The owner of this Dickensian establishment didn't seem to be particularly eager to sell anything. He sat in the back of the store, apparently paying little attention to the patrons.

When approached, he would peer over his glasses with an air of slight annoyance and deal imperiously with the intruder. In my case, he would steadfastly maintain that it was impossible that he had in stock any book that I didn't already own. On several occasions, I did find interesting books I didn't have yet, including a fine specimen of a famous field guide from the beginning of the century, Charles McIlvaine's *One Thousand American Fungi*. My dedication and my frequent visits broke through the gelid reception he usually offered to "customers," and after some time he and I had a game going. Still, no sooner would I set foot in the store than a barking voice could be heard from the back: "You already have everything I carry!" In a timorous way, I would acknowledge that, as long as I was there, shouldn't I just as well have a look around?

We kept up this game for a couple of years, both of us stubbornly sticking to our guns. One day I challenged the proprietor that if I could find in his stock a mushroom book that I didn't already own, he would have to sell it to me for a dollar. With a disdainful smile, he rose to the occasion and confidently agreed. Sure enough, my search of the shelves did not reveal anything new. So, I turned to the tables containing pamphlets and looked through a collection of assorted Americana. I found orations on the death of President Grant, advice on home canning of turnips, and nineteenth-century opinions on land policy in Ohio. Finally, after considerable search, I came up with a thin, blue-covered booklet, *Railroad Tie Decay*. This was a 1939 publication of the American Wood-Preservers' Association, complete with colored and black-and-white photographs of bracket fungi sprouting from the ends of railroad ties. One half of this booklet, entitled "Defects in Cross Ties, Caused by Fungi," had appeared earlier in the *Cross Tie Bulletin*. The price on the inside cover was three dollars. So, I won the bet. I admit that I relish owning this 54-page booklet as much as my more valuable books on mycology.

Leafing through this unique booklet, I found such politically incorrect headings as "What Fungi Should the Inspector Discriminate Against?" But it was easy to see why, from a railroader's perspective, fungi are the enemy. Who would want to see the wooden ties that hold the rails in place undergo "brown cubical rot"? One particular mushroom grows even on ties made from evergreens treated with creosote. This sturdy, gilled mushroom (*Lentinus lepideus*) is known as the "Train Wrecker." The name is not just a metaphor, for the damage *L. lepideus* causes to the ties as the fungal filaments "eat" the wood is often unsuspected. Some train derailments have been blamed on the decomposition of the ties caused by this fungus. The Train Wrecker belongs to the same genus as shiitake (*Lentinus edodes*)—or it did until recently, when taxonomists started to mess with it. By association, the U.S. Department of Agriculture became concerned with the potential damage that shiitake could cause to trees and deplorably declared a ban against its importation. Happily, shiitake is not particularly destructive and the U.S.D.A. has since relented.

"Dry rot" is a term that brings fear to the hearts of men. Any timbers left in contact with the ground or kept wet and exposed to the air will eventually succumb to the decomposing powers of fungi. The effects of this decomposition, of which dry rot is but one type, are real, and they can be frightening. The name is misleading, as wood "rots" only when moist or wet. But the name has stuck, because it is so descriptive: the rotted wood turns into dry-looking square blocks that eventually are reduced to a whitish powder. The main fungal species that causes dry rot is called *Serpula lacrymans*. This dramatic species name means "weeping," which does not refer so much to the condition to which it reduces owners of affected buildings or ships as to the water droplets that exude from the fungal surface when growth is abundant. Dry rot does not "fruit" into a mushroom but rather makes soft, fan-like sheets that can cover the surface of walls. Copious growths are sometimes found in abandoned houses where the wood has got wet from burst water pipes. Whole walls in such houses can be literally covered with the growth of the fungus, as though "The Blob" or some such extraterrestrial creature has invaded the premises.

Both home-owners and ship-owners used to be greatly concerned with the effects of dry rot, and it's only in recent times that suitable countermeasures (such as treating wood with effective preservatives) have become available. The threat to wooden ships was real, and many a naval battle was lost, or not even fought, because of the decay of the ships' hulls. In England, the trouble was caused by the scarcity of oak and by the use of unseasoned and untreated wood, with the consequence that naval ships were known to rot before they were even launched. Poor ventilation in the holds seems to have been a major contributory factor. It is likely that the sister ship of the *Mayflower,* the *Speedwell,* did not make the trip to the American colonies because it was suffering from "distemper of timber," as dry rot was then called. However, the need for chemical treatment of timbers has been known since antiquity. Noah was told: "Make thee an ark of gopher wood; room shalt thou make in the ark; and shalt pitch it within and without with pitch." Whatever "gopher wood" may be, the injunction to "pitch" was good advice.

Counterintuitive as it seems, extensive fungal colonization may be helpful to a host tree. Fungi do not usually invade the living sapwood, where the growth of the tree takes place; for the most part, they confine their spread to the dead heartwood, the central portion of the tree. Fungal growth leads to decay of the heartwood and, in time, to the hollowing out of the tree, turning it into an empty cylinder. A hollowed-out tree may actually become more resistant to high winds than a solid one, which is more unbending and heavier. In fact, very old trees rarely remain solid through and through. Ancient oaks, gnarled and hewn out inside, have been known to survive for 500 years or more, and some damaged-looking sequoias and redwoods are even older.

Rot can even be put to good use. Woodturners appreciate the discoloration of wood caused by the early stages of rotting, for pleasing color patterns may result. Dark lines in the wood make random and seemingly unpredictable designs well suited for bowls or vases. "Spalted" wood is a challenge to the woodturner because it is composed of soft and hard zones close together, making it hard to work on a lathe. The dark lines are due to the growth of fungal filaments and the chemical changes they produce. The trick for the wood-worker is to find a log in the right condition, because if rotting has proceeded too far the wood is unworkable. Mush-rooms growing from a felled or dead standing tree are a clue that the wood may be spalted. Maples, birches, and beeches seem to produce the best spalted wood. In England, other fungus-infested wood was put to a similar use. The wood of trees sometimes turns bright bluish-green from the growth of a fungus called *Chlorociboria aeruginascens*. It was used to make a variegated veneer that was applied to boxes and tables, a style known as Tunbridge ware.

Mushrooms contribute in an even more important way to the well-being of trees. The underground filaments of many fungi colonize the roots of trees to the clear benefit of the host. When fungal filaments come in contact with the ter-minal tree roots, they cover them with a thick layer known as the mantle. Rootlets of such trees look swollen and club-shaped and are clearly seen to be surrounded by the mantle of fungal filaments. These rootlets lose the hairs they nor-mally use for absorbing water and minerals and now depend entirely on the fungal filaments to provide these nutrients. This arrangement extends the absorptive capacity of the tree from that of its roots to that of the vast underground network of fungal filaments, greatly increasing the amount of material that can be taken up from the ground. For the tree, the fungal filaments become enormously long and far-flung drinking straws.

This root-fungus relationship is known as *mycorrhiza* ("fungus-root"). Trees with mycorrhizal associations grow better than those lacking them, as becomes evident when an area is reforested with trees not native to the area and unaffected by the local fungi. Reforesting efforts achieve their greatest success when the trees are artificially seeded with cultures of fungi capable of establishing mycorrhizae. Cultures of mycorrhizal fungi are sold commercially for the purpose of inoculating planted tree seedlings. In some flowering plants that lack chlorophyll and cannot carry out photosynthesis, such as the ghostly pale Indian pipe (*Monotropa uniflora*) or the bright red snow plant (*Sarcodes sanguinea*), a mycorrhizal association is not just beneficial, it is obligatory. These plants could not survive without their mycorrhizal fungi, which supply them with the sugars, nitrogen, phosphorus, and micronutrients needed for growth and survival without chlorophyll. So much for judging fungi to be sordid and depraved entities.

Sooner or later, most trees end up decorated. Protruding shelves and brackets of various sizes, shapes, and colors will bedeck just about any dead tree. Large or small, bracket fungi seem to become one with the old decaying tree. Sometimes they are called "conks," because they mimic conch shells in shape. Most are distinctly woody or leathery, some so tough that they can only be dislodged with a hatchet or a saw. They vary in color and shape, but most make flat, semicircular projections, whereas some look like horse hooves sticking out on the side of a tree.

Bracket or shelf fungi are members of the polypore family. The underside of a typical polypore consists, unsurprisingly, of innumerable pores. It is similar to the underside of a bolete cap, but polypores are much too woody to be confused with boletes. Also, the boletes have individual tubes that can be separated from the rest of the tissue, whereas the polypores do not. The pores are usually round but some are honey-combed or elongated. There are even polypores that have what looks like gills instead of pores. A hand lens comes in handy and reveals a lovely set of geometric variations that can escape examination with the naked eye.

Bracket fungi are an extensive group that vary in size, shape, and habitat. The specimens of some species are huge, not infrequently forty inches across. The current size record for the fruiting body of any mushroom is probably held by a specimen of *Bridgeoporus nobilissimus* found in 1943 in the state of Washington. It measured four-and-a-half feet by three feet and weighed three hundred pounds. So few specimens of this species have been found that it became the first fungus to be listed as an endangered species in the United States. Other polypores are small and are mainly noticed because they grow abundantly, covering dead branches or trunks of diseased trees. A few are ground-dwelling. Many of the larger bracket fungi are perennials. They make a new layer of pores each year, their age limited by the number of years the dead tree remains erect. By sawing through one venerable polypore specimen, I once counted fifty-two layers of pores, one on top of the other.

Some species of polypores are shapely and have lively colors, enough so to be used as jewelry, wall ornaments, or decorations in flower arrangements. Many polypores keep well after drying, retaining both color and shape for many years. Particularly pretty are the turkey tails (*Trametes versicolor*), little shelves about an inch across with concentric zones of different brown and tawny colors. They have fine hairs on their surface and feel velvety to the touch. I enjoy them as winter friends and count among my pleasures seeing them on snow-covered trees as I glide by (somewhat unsteadily) on cross-country skis.

Polypores have many uses. They were used to make "amadou," a form of tinder that was widely known in Europe until matches and other convenient ways of making fire became readily available. Amadou, also known as "German tinder," readily caught fire when hit by the spark from a flint. To make amadou, a common bracket fungus (*Fomes fomentarius*) was pounded into fibers, dipped in a solution of saltpeter (potassium nitrate), and allowed to dry. Amadou had a myriad of other uses, too: to stanch blood from the cavity of a tooth after an extraction; to serve as padding over sensitive parts of the body; even to stuff hats designed to pre-vent baldness. As mentioned before, the prehistoric hunter found frozen in the Austrian-Italian Alps carried among his paraphernalia what was probably a kind of amadou. Poly-pores have also been used in many parts of the world as portable embers, to keep fires going. Smoldering in the mid-dle of a large polypore, a small fire will keep burning for hours. The fungus and its embers can readily be carried around and may serve not only as a source of fire but also as a portable heater. On a trip to Alaska, my wife and I were shown around a native fishing camp on the Tanana River, not far from Fairbanks. The guide, a young woman with Atabascan roots, held up a specimen of *Fomes fomen-tarius* and explained its ancient use for keeping fires going. She called what she was holding a "growth from a tree," which is probably how her ancestors described the fungus. The same view is held by the Cree Indians in Northern Quebec, who also use the fungus embers as a portable source of fire.

One especially handy species (*Piptoporus betulinum*) has been employed as a strop for sharpening razors, as a silver polisher, and, in a pinch, as imitation potato chips. I have tried *P. betulinum* on a tarnished silver tray and found that it works quite well, albeit too slowly to make the tedious chore any more pleasant. Brackets of this species grow to ten inches or more across and are invariably found on dying white birches wherever these occur in North America. Other polypores have been used as combs for horses or, when hollowed out, as serviceable flower pots. They have even been carved as ceremonial war maces in New Guinea.

Within the diverse repertoire of the polypores, the genus *Ganoderma* stands out. One species of this group is known as the artist's fungus (*Ganoderma applanatum*) and is often found growing on dying oaks and sometimes other trees. This fungus has the shape of a semicircular shelf and is brown with broad and knobby concentric ridges on the top. It can be two feet across or even larger. It merits its name because its underside, which is light in color, turns a dark brown when scratched. Fine drawings can be made on this surface with a thin needle or knife. In the hands of a skilled artist, the work can resemble scrimshaw carved on ivory. Dried specimens of the artist's fungus keep well, and some people display them on the mantle. One particular specimen of this polypore has even become an object of veneration. A church in the Mexican state of Puebla, Nuestra Señora del Honguito, is dedicated to a fragment of *Ganoderma* showing what is believed to be the image of Christ on the cross.

A cousin of this fungus is both spectacular-looking and, according to folklore, effective as a cure of a variety of ills. The half-circular brackets of *Ganoderma lucidum* have a shining layer of bright vermilion on the top surface, often the same color and brightness as a Japanese lacquer box. It is difficult not to believe that someone came through the woods and covered them with varnish or shellac. This species has been used in East Asia for the treatment of cancer and other diseases. The mushroom is tough and cannot be consumed intact, so it is ground up and soaked to produce a tea, or it is made into candies. It is known in Chinese as *ling-zhi* and in Japanese as *reishi,* names that have been translated as "marvelous vegetable" or, more dramatically, if inaccurately, as the "mushroom of immortality." It is not uncommon to find a good-sized, dusty specimen in the window of a Chinese herbal drug store.

Surprising as it may seem (until you think of it), mushrooms have been used for dyeing textiles. Miriam Rice has written a book on mushroom dyes with illustrator Dorothy Beebe (see "Resources"). Certain species of polypores or of fleshy mushrooms can be used to make quite intense dyes that are used in conjunction with mineral salts (mordants) that help fix the dye to the textile fiber. Depending on the mordant used, the same mushroom can be used to dye the same material with different colors. For example, a Pacific coast mushroom of the *Cortinarius* group (*Dermocybe phoenicea* var. *occidentalis*) produces reds, blues, purples, or grays. A common polypore, appropriately called the dyer's polypore *(Phaeolus schweinitzii),* produces various rich hues of yellow, orange, reddish, and brown. In addition, Miriam Rice has found a way of making fine paper out of mushrooms.

The appeal of the polypores extends to gastronomic uses. Most polypores are too tough to eat but there are two notable exceptions, both with poultry names. They are the hen-of-the-woods *(Grifola frondosa)* and the chicken mushroom

(Laetiporus sulphureus), which were described in Chapter 7. Finding either of these mushrooms is like hitting a jackpot because they usually make large specimens, twenty pounds or more in weight. Other polypores are also considered edible, but usually only when they are very young. Even the tougher ones can be broken into pieces and used to make a tasty, mushroom-flavored consommé after lengthy cooking.

I find it appropriate to have left the polypores for later in the book. Because of their greater longevity, they represent the world of fungi in a way more enduring or seemingly more trustworthy than that of their ephemeral cap-and-stem relatives. To me, they seem less implausible than the other members of the fungal kingdom. They fit more readily in the scheme of the woods. I think of the polypores, adorning the trees at all times of the year, as being innate to the forest.

13

INSECTS AS FUNGUS GARDENERS

The landscape of the savanna of sub-Saharan Africa is conspicuously dotted with *termitaria,* huge earthen mounds built as nests by termites. These cone-shaped structures get to be as high as eighteen feet and as wide as ten feet at the base, and have a curious resemblance to something manmade, like a concrete tower constructed too hastily. The traveler unfamiliar with that region of the world must wonder what their purpose may be. Termitaria are also found in the forests of India and other parts of South Asia, but the Asian nests are not as prominent as the African ones, being smaller and partly obscured by vegetation.

The packed earth of a termitarium is criss-crossed by tunnels and chambers in which the termites live and, to put it in anthropomorphic terms, cultivate their fields to grow their food. A complex system of flues and chimneys keeps the nest at the right temperature and humidity. The termites that build these impressive structures are social insects with most impressive skills as both engineers and farmers. As

233

many as a million individuals cooperate in the construction and maintenance of one of the larger nests. These populations require large amounts of vegetation, including wood and plant debris, as food. Some termite species prefer wood already decayed by the action of fungi. In the tropics of the Old World, termites compete successfully with human farmers in that they remove organic litter from the ground and thus impoverish the soil. Termites are no more welcome by African and Asian farmers than they are by American homeowners.

Although termites have earned their reputation as pests for their ability to destroy wood and other plant material, they cannot digest lignin and cellulose directly. Instead, they use one of two general strategies to break down these indigestible compounds. The termites found in the temperate zone, the scourge of the home-owner, carry bacteria and protozoa in their gut that are able to convert wood into palatable sugars. Much like cows and other ruminants, these termites depend on microscopic partners in their digestive tract for their nutrition. Incidentally, cows and termites share another attribute: the production of large amounts of methane. How can these vastly different life forms share such an unusual metabolism? It is because the complex microbiology of cellulose decomposition in the gut of both facilitates the growth of a type of bacterium that produces methane.

The termitaria-building termites use a completely different strategy for cellulose degradation; fungi external to the insect's body digest wood and other vegetable matter for the termite. In a way, the fungi are a portable gut. The use of fungi by these insects is far from haphazard: the termites assiduously cultivate the fungi. In startling analogy with how humans raise mushrooms, the termites grow "fungus gardens," fields of cultivated fungi housed in miniature caves or galleries inside the termitaria.

As with all social insects, the termites are divided into castes—such as the workers, the soldiers, the king and queen —each programmed to perform distinct duties. The workers eat wood and other forms of vegetation and carry to the nest this material in their gut, digesting some of it as they go. Once they reach a suitable place in the termitarium, the workers excrete the partially digested material and thump it down on the surface. In the termite gut, the mash had previously become permeated with filaments of a fungus of the *Termitomyces* genus. Once excreted, the fungal mycelium grows into tiny spheres, about the size of a small pinhead. These spheres, packed with fungal spores, are the most prominent feature of the fungus gardens. To the termites, the scene must appear as a field of tightly packed giant puffballs would to us. The worker termites graze on the spheres, which are their principal source of food. The workers are the only caste able to eat directly and they in turn must feed the larvae as well as the members of all the other castes. The queen of the colony spends most of her life sealed in a royal chamber and obviously cannot move about in search of food. The soldiers cannot eat because their gigantic mandibles, used for weapons, get in the way.

The agricultural know-how of these lowly creatures is so intricate that comparisons with human farming come readily to mind. Termites perform all the major tasks of proficient farmers: preparing the soil, fertilizing, pruning, weeding, and harvesting. They must do a great job of weeding, because a functioning fungus garden usually consists of a single species of fungus, a feat that trained mycologists sometimes have a hard time achieving in the laboratory. Unplanned fungi are "weeded out," perhaps by chemical fungicides produced by the termites, perhaps by mechanical removal. When a nest is abandoned, it quickly becomes overgrown by other fungi, indicating that the termites knew how to discriminate between "wanted" and "unwanted" fungi while tending their gardens. The effect of pruning or cropping is best seen by comparing a well-tended fungus garden and the same fungus growing in a laboratory culture. In artificial culture the fungi produce long filaments, whereas under the care of the termites the tips of the fungi become rounded into clubs and spheres, shapes not seen elsewhere. Abandoning a garden results in the outgrowth of the fungi: they "go to seed."

Eventually, when the termites die or leave the nest, the fungi develop into large, beautiful mushrooms that stick out from the surface of the termitarium. Mushrooms, in other words, arise from the lack of proper pruning. The genus *Termitomyces* includes a couple dozen varieties of mushroom that are relished by the local people. These are among the most prized mushrooms in the markets of Nigeria, Zambia, and the tropical countries of southern Asia. One is tempted to ascribe a particularly fine palate to the termites, given that their ordinary food is what we humans regard as an exceptional treat. Some *Termitomyces* mushrooms reach stupendous sizes. The appropriately named *T. titanicus* has produced what may be the largest fleshy fungus known, with caps up to forty inches in diameter. Caps twenty inches across and weighing over four pounds are not uncommon, although most harvested specimens are less than twelve inches across.

In tropical and subtropical America, the role of the termites and their fungi is taken over by the leaf-cutting ants, also known as the parasol ants. Found from Argentina to the southern United States, these ants also subsist solely on the fungi they grow. So large is their number that they constitute the most abundant invertebrates in tropical America. Fungus growing is big business indeed.

Certain leaf-cutting ants, such as *Atta texana,* make huge *underground* nests, to which they bring pieces of leaves, flowers, and other organic materials to construct elaborate fungus gardens. In cross-section, the fungus gardens appear as whitish, irregular masses that almost fill the large cavity of the nest. Characteristically, the cavities made by *Atta texana* are about a foot across and are located some three feet below ground, connected by tunnels to the surface. The various agricultural chores are here also divided among members of the different castes. The larger workers gather sections of leaves, often half an inch across, and carry them in long and fast-moving processions to their nest. The ants effortlessly carry a burden that to a human would be as heavy and awkward to carry as a large plywood panel. Given this industrious dedication to gathering leaves, it is not surprising that early observers thought that the leaves themselves were the food for the ants. It took considerable investigation to find out that the plant material only serves as a growth medium for the real foodstuff, the fungi. The person who first suspected this was an Englishman, Thomas Belt, working in Nicaragua in 1874. He felt compelled to state that this explanation was "extraordinary and unexpected." Confirmation came from experiments carried out in Brazil by Alfred Möller and published in 1892.

The ants cut off the pieces of leaf by slicing with their scalpel-like mandibles. This may seem to be a difficult thing to do, as can be attested by anyone who has tried to cut a loosely held piece of paper by slicing it with a knife. A German entomologist, Jürgen Tautz, found that the ants solve the problem by emitting vibrations that effectively steady the leaf and make it easy to cut. Tautz found that when he mimicked these vibrations artificially, the leaf moved back and forth over such a small distance that, practically speaking, it remained rigidly in place. The vibrations have a frequency of about 1,000 per second, and they are audible to people as a faint humming sound. The vibrations are also heard by other ants, who are attracted to the same leaf to share in the dismantling process.

Some species of leaf-cutting ants leave trails of denuded earth between their nest and the source of vegetation. These clear trails can reach astonishing dimensions, over six hundred feet in length and twelve inches in width. The amount of coming and going is remarkable, often resulting in traffic jams where the outbound and inbound individuals must literally climb on top of one another (a technique that has yet to be developed for the cars on California freeways). This show is well worth watching, although the observer would be wise to keep a prudent distance because the larger workers and the soldiers can inflict painful wounds. In fact, the soldiers have such powerful jaws that they have been used for suturing or, more accurately, stapling wounds. As early as 1653, Bernabé Cobo, in his *Historia del Nuevo Mundo,* described how Indians in Santo Domingo would bring the two edges of a cut together and have certain kinds of ants bite across it. They then "cut off the insects' heads, which remain attached to the wound with their mouths or mandibles as firmly closed as they were in life."

Displaying extreme efficiency, the ants prepare the leaves for composting even before they reach the nest. Small workers go out on the foraging expeditions and return as passengers on the pieces that are being carried back by the larger workers. On the way, the small ants swab the leaves with their saliva in preparation for further handling. They cling tenaciously to the leaf fragment and keep on licking, no matter how rough the ride. At the nest, the larger workers carry out an elaborate composting operation. The leafy pieces are cut into smaller sizes, licked thoroughly, and mixed with fecal material. Using their mandibles, legs, and antennae, the ants then fold these fragments over and over, until they can knead them into tiny, juicy balls. This is done with enormous care: an individual ant may take as much as a quarter of an hour to mold a single pellet into its proper shape. This pulpy material is carefully deposited at the edge of the garden and jabbed into place. It is then "seeded" with the mycelium from the older sections of the garden. The garden rapidly becomes permeated with new filaments. The fungal surface consists of aggregates of hyphal tips that end in roundish bodies, as the fungi in termitaria do. Here, too, special shapes are characteristic of the fungi growing in association with insects. When the ants quit tending their garden, the fungi grow into long filaments. Unlike the fungi in termitaria, however, these filaments seldom develop into conspicuous mushrooms.

Leaf-cutting ants have received very bad press because of the damage they cause to vegetation and the threat they constitute to cultivated crops, such as coffee and cacao. They are, in fact, the dominant herbivores in the American tropics. In Brazil, the *saúva,* the local name for these ants, inspired an old saying: "Either Brazil gets rid of the saúva, or the saúva gets rid of Brazil." In partial defense of the ants, it should be mentioned that they do not completely defoliate the plants on which they feed. In addition, the spent litter from the fungus gardens, still very rich in organic matter, is returned to the environment, where it serves as plant fertilizer. Typically, this material is carried up the trunk of a tree or even over vines and allowed to drop to the ground. The ants of one nest have been observed to carry the waste to a smooth rock and allow it to tumble down the slope.

Luckily for the sake of our understanding, the ant-fungus relationship has drawn the attention of researchers. It has been shown, for example, that the relationship may be remarkably old, dating perhaps to fifty million years ago. When the more specialized groups of ants selected a given fungus as a favorite crop, they remained with it for eons. This means that the fungi used by a given species of ants are descendants of the same spores. The more primitive species, on the other hand, are less specialized and seem to be able to sample new kinds of fungi. These conclusions are drawn from studies by researchers from the U.S. Department of Agriculture, the Marine Biological Laboratory at Woods Hole, Massachusetts, and Cornell and Harvard universities. Comparing the so-called ribosomal RNA (a specific kind of nucleic acid) of different species of ants and their fungi, the researchers learned that once the symbiotic relationship was established, the two partners remained true to each other over the eons. In a way, this longevity is not surprising: as Harvard's eminent entomologist E. O. Wilson pointed out, this adaptation to a most efficient utilization of fresh vegetation is so unusual and successful that "it can properly be called one of the major breakthroughs in animal evolution."

Termites and ants are not the only fungus-eating insects. Ambrosia beetles, a kind of wood-boring beetle, also depend on their fungi for sustenance. These small insects bore tunnels in the sapwood, sometimes of healthy young trees. Certain species make tunnels that branch into a complex system of galleries. These insects do not usually kill the trees but spoil the wood for commercial purposes. They therefore represent a serious problem for forest managers both in the tropics and in the temperate zone.

The beetles sow the walls of the tunnels with spores they carry in specially designed pouches called "mycangia." The spores germinate and produce filaments that radiate into the wood and extract nutrients from the host tree. A specialized fungal growth arises on the walls of the tunnels, making a glistening mat known as "ambrosia," the Food of the Gods. A pioneer student of these insects, H. G. Hubbard, wrote in 1897: "All the growing parts of the fungus are extremely succulent and tender . . . The young larvae nip off the tender tips as calves crop the heads of clover, but the older larvae and the adult beetles eat the whole structure down to the base, from which it soon springs up afresh, appearing in white tesselations upon the walls . . . The growth of ambrosia may be compared to asparagus, which remains succulent and edible only when continually cropped, but if allowed to go to seed is no longer useful as food." The fungi provide the beetle not only with all nutrients but also with the steroids needed to make the hormones that regulate the beetle's metamorphosis, a most unusual relationship between nutrition and development.

Ambrosia beetles are also sophisticated mycologists. They readily distinguish their favorite species of fungus from others. A given species of beetle has a distinct preference for a particular species of fungi, although it can subsist on others. At first glance, it may seem that their association with the beetles is not of much benefit for the fungi. Note, however, that the fungi eaten by beetles have a greater chance for growth and, even more important, for being dispersed to other sites. The fungi belong to various genera of molds or yeast and bear such names as *Ambrosiella, Ambrosiomyces,* and *Ceratocystis* (the genus that includes the Dutch elm disease fungus). Ambrosia beetles are found in many parts of the world and are abundant throughout North America.

Variations on the theme of "fungus farming" by insects abound. Some midges deposit their eggs in plant tissues, which then develop into round, leathery "galls" on the plant stem. The inside cavity of the gall becomes covered with mycelium which becomes the food for the emerging larvae. By inoculating the plant with spores at the time of egg laying, the female midge ensures that her offspring will be provided with their mycological birthright. Dependence on fungi by the young is also a characteristic of certain wood wasps, whose larvae obtain their food from bracket fungi.

In some cases the roles are reversed and the insect provides food for the fungus. For example, a fungus called *Septobasidium* has specialized cell structures (haustoria) that penetrate the bodies of certain scale insects to extract nutrients from their blood. The insects, in turn, obtain food from plant sap by means of long sucking tubes. Even dead, the insect continues to function as a long drinking straw, in effect becoming a conduit between the primary food source and the fungus. The individual insect may die, but its brethren enjoy the protection of a fungal roof over their heads to shield them from predators. Yet another group of fungi, ascomycetes of the order Laboulbeniales, live on the chitinous outer shells of living insects without apparently harming the insects in any way. It is not clear whether they just go along for the ride or interact with the insects in some other manner.

   Insects and fungi do not encounter one another by chance but have evolved ways to make sure that partnerships between them are continuous. Were this not the case, they would have to reestablish relationships at every generation. How do fungus-gardening insects ensure that new colonies or individual offspring come into close contact with the beneficial fungus? Some insects have special organs in which they carry fungal "seed" material; for instance, the mycangia of the ambrosia beetles are filled with spores to be used for inoculating the tunnel walls. Likewise, the virgin queen ants and queen termites fill special body cavities with pieces of the fungus garden before embarking on the nuptial flight that will lead to the establishment of a new colony. As soon as they burrow to make a new nest, they start immediately to build a new fungus garden. To ensure success in the early stages of the garden, the queen ant fertilizes it with fecal liquid, a task taken over by the workers later on.

Who gains from all this—the insect or the fungus? Clearly the insect is better off for eating the palatable fungus rather than indigestible wood or other plant material. The fungus, in turn, benefits because the insect works to promote its growth and spread. Some species of fungus, such as the *Termitomyces* species, are not found in nature *except* in association with termites. The answer, then, seems to be that both partners benefit from this curious relationship, often to the point of not being able to live without the other. This may seem strange at first glance, but a profound dependence on fungi is something that mycologists can certainly appreciate.

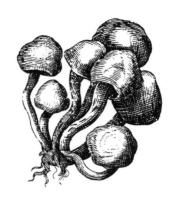

EPILOGUE ~ THE BIOLOGIST AS MUSHROOM HUNTER

During a vacation taken while I was just putting the finishing touches on this book, my wife and I were walking a trail near Oregon's Crater Lake. She pointed out to me a small collection of mushrooms that were growing out of a shallow snowbank. Her ability to spot mushrooms is legendary in our household, but finding them in this habitat seemed unusual. The specimens were respectable in size, about an inch across the cap and two inches tall, rosy pink in color, and covered with a shiny layer of slimy material. Baby mushrooms were still entirely encased in the snow. I identified the specimens as belonging to the wax caps *(Hygrophorus goetzii)*, a species that is one of the few known to grow in snowbanks.

The realization that life can be sustained at unexpected extremes of temperature has generated much excitement among biologists. Life at high temperatures has generally attracted more attention than that in the cold. Bacteria have been found growing at temperatures as high as 113°C, about 23°F above the boiling temperature of water at sea level. Obviously, for water to reach such high temperatures without boiling, it must be under high pressure, such as is found in the depths of the ocean. How can life exist, and even thrive, at such extreme conditions?

Few answers are available to date. It turns out, however, that these "extremophiles" are not just a biological curiosity. They have industrial applications as well. The enzymes of these organisms are themselves resistant to high temperatures and can be expected to be employed in a variety of technological processes. One of these enzymes is used in the reaction that amplifies DNA, known as "polymerase chain reaction," or PCR, a technique that got a lot of press during the O. J. Simpson trial. This particular enzyme, a DNA polymerase, was derived from a bacterium isolated from a hot spring at Yellowstone Park.

The tenacity of life forms is just as surprising in the cold. Fungi, mainly microscopic molds and yeast, can grow at temperatures well below freezing, which is why food must be preserved at very low temperatures. How do these organisms thrive, given that growth is not possible in the frozen state? The answer is that they are not frozen themselves, despite the low temperature of their environment. These cold-loving life forms keep from freezing by generating some heat through their metabolism and by making chemical antifreezes or taking up high concentrations of salts. This strategy is also used by fish that swim in seawater below the usual freezing point, 0°C.

Our view of what we call extreme conditions is anthropocentric. Any circumstance that we ourselves cannot tolerate is considered extreme. It takes bacteria, mushrooms, and other fungi to show us that this is a parochial view of life.

On the hike with my wife, I had no way to measure the highest temperature to which the mushrooms were exposed in the snowbank. It may well have risen above freezing in their vicinity, at least for short times. However, cold these specimens were. Perhaps the slimy layer that surrounded the specimens served as their antifreeze. It may not be a coincidence that many mushrooms that grow at low temperatures are in fact slimy. Whatever the mechanism, ours was an unexpected sight. I felt reassured that, given such prowess at survival in an unfriendly environment, mushrooms are probably here to stay.

Just as I started it, I end this book on a personal note. I have known all along that the subject of mushrooms is a rich one, as our discovery of the mushrooms in the snowbank illustrates, but I have nevertheless been surprised as this book unfolded that there is so much to write about it. As I was working, more and more interesting material kept appearing. Sometimes this was in the form of stories, either discovered in books or told to me; sometimes it was experiences of my own that came to mind. I ended up reaping a greater harvest than I had expected, for which I am obviously glad.

I am a microbiologist by profession, now near the completion of my career. When I collect and study mushrooms, however, I do not act as a professional biologist. Most of what I love about mushrooms and how they fit in people's lives is far remote from my research and teaching. My life in science has been spent at the laboratory bench: I study how bacteria make their DNA as they grow. I am of the generation that witnessed the beginning of molecular biology and its offspring, genetic engineering. It was only after my career was established that I stepped into the world of forests, pastures, and mushrooms.

The distance between these two interests—microbiology and mushrooms—may not appear to be very fundamental to a non-scientist, but it is in fact quite considerable. It is true that biology is biology, in the sense that the basic question is always the same—What is life?—but at the level at which we participate in the *profession* of biology there are marked differences in attitude between those who study living things in the field and those who work in laboratories.

This gap is of recent origin—it was unknown until the nineteenth century—and, happily, it gives signs of closing. On the one hand, biologists who study the evolution of living things and their place in the environment are coming into the laboratory to take advantage of modern molecular tools. On the other hand, those who study the functions of living cells have found great opportunities in probing the wondrous diversity of the natural world. The rift between the field worker and the lab worker, in the questions they ask and the attitudes they convey, is narrowing, and we can welcome the fact that biology is reemerging as a unified science. It is worth noting that the "history" in "natural history" is derived from the Greek word for "learning by inquiry," which today we would name "science."

For most of my professional life, however, the distinction between "field biology" and "laboratory biology" was quite substantial. My colleagues seemed content to study one or two kinds of bacteria under laboratory conditions and only rarely seemed concerned with the "real world." The view has been put forth that different personalities are attracted to the two approaches to biology. To overstate the point, the "naturalists" are seen as more caring, more accepting of their role as stewards of living things, whereas the "experimentalists" are thought to be more analytical, interested only in how things work.

That's the theory, at least. I have always had a hard time with this notion because it seemed, at best, to describe people at the extremes. I feel that I straddled these two worlds. Strange as it may sound, I have developed, if not a love, at least a personal closeness to the bacteria I study. The strains I have worked with are, by and large, harmless to people. To me, they are living things, not just bags of enzymes and DNA. They are, in other words, as alive to me as mushrooms are—and as trees and animals are. So, is the jump from the lab bench to the woodland glade as big as all that? In both places one can find, or make, the opportunity to study nature, to experience life, and it is my hope that this book will lead you at least part of the way toward that end.

RESOURCES

CREDITS

INDEX

RESOURCES

An Annotated Bibliography

Mushrooms and Other Fungi

Ainsworth, G. C. 1976. *Introduction to the History of Mycology*. New York: Cambridge University Press. Not only is this the definitive history of mycology, it also encompasses many other interesting aspects of the fungi. It is particularly well illustrated.

Brodie, H. J. 1975. *The Bird's Nest Fungi*. Toronto: University of Toronto Press. The definitive work on the subject by its leading investigator.

———— 1978. *Fungi, Delight of Curiosity*. Toronto: University of Toronto Press. Vignettes of some biological phenomena exhibited by common and unusual fungi are presented in a personal style.

Buller, A. H. R. 1909–1934. *Researches on Fungi,* volumes 1–6. London: Longmans, Green. This work includes classic reports of some of the most illuminating studies on spore dispersal, especially in volumes 1, 2, and 4. Available in specialized libraries only.

Findlay, W. P. K. 1982. *Fungi: Folklore, Fiction and Fact*. Eureka, CA: Mad River Press. This concise, fact-filled book deals mainly with mushrooms in history.

Hudson, H. J. 1986. *Fungal Biology*. New York: Cambridge University

Press. A remarkably well-written scientific book on fungal ecology, this book serves professionals and amateurs alike. The latter should be warned that some chapters deal with such uncelebrated subjects as "Fungi as decomposers of leaves" and "Fungi as inhabitants of animal feces."

Ingold, C. T. 1971. *Fungal Spores: Their Liberation and Dispersal.* Oxford: Clarendon Press. A technical but lively account of the myriad strategies used by fungi to disperse their spores.

Keewaydinoquay. 1978. *Puhpohwee for the People.* Cambridge, MA: Botanical Museum, Harvard University Ethnomycological Studies, No. 5. Written by an Ojibway herbalist and shaman, this booklet describes various uses of mushrooms by Native Americans and includes some lively narratives.

Kendrick, B. 1992. *The Fifth Kingdom.* Waterloo, ON: Mycologue Publications. This basic textbook, written in a pleasant, often amusing style, is a fine introduction to many aspects of mycology.

Parker, L., and D. T. Jenkins. 1979. *Mushrooms: A Separate Kingdom.* Birmingham, AL: Oxmoor House. This general introduction to the mushrooms is illustrated with many beautiful watercolors and a text rendered in lovely calligraphy.

Ramsbottom, J. 1972. *Mushrooms and Toadstools.* London: Collins. This is a treasure trove of information on the nature of mushrooms and their role in human events through history.

Wasson, V. P., and R. G. Wasson. 1957. *Mushrooms, Russia and History.* New York: Pantheon Books. The most comprehensive and challenging compilation of the interactions of mushrooms and humans, this book was issued in an expensive, hard-to-find edition. It may be found in some of the better-endowed libraries.

Collecting, in Solitude and in Groups

Note: Regional field guides are also available for many parts of the United States and Canada. Some of these are distributed by major publishers and may be relatively easy to obtain, others are issued by smaller houses and may not be readily available at bookstores. Seek the advice of local mycologists to find the best regional guides.

Arora, D. 1986. *Mushrooms Demystified,* 2d ed. Berkeley: Ten Speed Press. Written in a comprehensible and witty style, this is a robust book that serves the needs of the serious amateur.

———— 1991. *All That Rain Promises and Much More . . .* Berkeley: Ten Speed Press. A smaller and even livelier version of the above, this booklet is meant for beginners.

Fisher, D. W., and A. E. Bessette. 1992. *Edible Wild Mushrooms of North America.* Austin: University of Texas Press. A practical "field-to-kitchen guide" that includes nicely illustrated recipes.

Friedman, S. A. 1986. *Celebrating the Wild Mushroom.* New York: Dodd, Meade. As the title suggests, this is a personal narration of the joys of wild mushroom hunting and eating.

Lincoff, G. 1981. *The Audubon Society Field Guide to North American Mushrooms.* New York: Alfred Knopf. This authoritative field guide fits readily in everyone's basket. The descriptions of mushrooms are particularly "user-friendly."

Marteka, V. 1980. *Mushrooms: Wild and Edible.* New York: Norton. Not a field guide in the strict sense, this book combines engaging descriptions of common edible and poisonous mushrooms with recipes and discussion of preservation and other useful subjects.

McKnight, K. H., and V. B. McKnight. 1987. *Mushrooms.* Boston: Houghton Mifflin. Included in the Peterson Field Guides series, this book has brief but well-written descriptions of many of the mushrooms of North America and beautiful watercolors.

Phillips, R. 1991. *Mushrooms of North America.* Boston: Little, Brown. A collection of fine color photographs, this book is especially useful for those who enjoy attempting to identify mushrooms by browsing through pages and pages of pictures.

Culinary Tales

Bessette, A. R., and A. E. Bessette. 1993. *Taming the Wild Mushroom: A Culinary Guide to Market Foraging*. Austin: University of Texas Press. A guide to cooking with the "exotic" mushrooms, it includes the kinds available in some supermarkets as well as those found in the woods and fields.

Carluccio, A. 1989. *A Passion for Mushrooms*. Topsfield, MA: Salem House. The author's enthusiasm for eating wild mushrooms pops up on every page and in every recipe.

Czarnecki, J. 1986. *Joe's Book of Mushroom Cookery*. New York: Athenaeum. Here are more than 300 recipes by the third-generation owner of a famous wild mushroom restaurant in Reading, Pennsylvania.

Freedman, L. 1989. *Wild about Mushrooms*. Reading, MA: Addison-Wesley. A collection of recipes arranged by the kind of mushroom, this book includes general information about mushroom cooking and preserving.

Grigson, Jane. 1975. *The Mushroom Feast*. New York: Alfred Knopf. A classic by one of the great cookbook writers of our time, this volume is especially suited for the gourmet cook. Available in paperback.

Hall, I., G. Brown, and J. Byars. 1994. *The Black Truffle: Its History, Uses and Cultivation*. Christchurch: New Zealand Institute for Crop and Food Research, Ltd. (Private Bag 4704, Christchurch, New Zealand). The biology and cultivation of the French black truffle are described in detail and elegant style by two New Zealanders and an American. Recipes are included.

Harris, B. 1986. *Growing Shiitake Commercially: A Practical Manual for Production of Japanese Forest Mushrooms*. Madison, WI: Science Tech Publishers. This book presents a detailed treatment of the cultivation of shiitake, on a small as well as on a larger scale.

Leach, P., and A. Mikkelsen. 1986. *Malfred Ferndock's Morel Cook-*

book. Dennison, MN: Ferndock Publishing (Box 86, Dennison, MN 55018). A collection of whimsical stories about morels and people, with recipes.

Leibenstein, M. 1986. *The Edible Mushroom: A Gourmet Cook's Guide.* Old Saybrook, CT: Globe Pequot Press. This small tome is a delightful compendium of mushroom history, lore, and interesting recipes.

Miller, H. 1993. *Hope's Mushroom Cookbook.* Eureka, CA: Mad River Press (141 Carter Lane, Eureka, CA 95503). Over 300 recipes are presented by one of the most experienced mushroom cooks in the United States.

Ratzloff, J. 1990. *The Morel Mushroom.* Stillwater, MN: Voyageur Press. This nicely illustrated booklet contains some unusual morel recipes, including one for the "Official Minnesota State Dinner."

Revised Code of Washington. 1989. *Wild Mushroom Harvesting and Processing.* Chapter 15.90. Here are the statutes that regulate commercial mushroom harvesting in the state of Washington.

Stamets, P. 1993. *Growing Gourmet and Medicinal Mushrooms.* Berkeley: Ten Speed Press. This highly informative, well-presented treatise explains the steps in cultivating a large number of mushrooms.

Stamets, P., and J. S. Chilton. 1983. *The Mushroom Cultivator.* Olympia, WA: Agarikon Press. The emphasis in this thorough book is on cultivation of the white button mushroom, but other mushrooms are presented as well.

Weber, N. S. 1988. *A Morel Hunter's Companion.* Lansing, MI: Two Peninsula Press. This guide to morel hunting includes much useful information on the taxonomy and morphology of morels and their relatives.

A Kingdom of Versatile Parts

Ammirati, J. F., J. A. Traquair, and P. A. Horgen. 1985. *Poisonous Mushrooms of the Northern United States and Canada.* Minneapolis:

University of Minnesota Press. A compendium of poisonous mush-rooms with keys to the identification of selected species.

Batra, L. R. 1979. *Insect-Fungus Symbiosis.* New York: Wiley. The pro-ceedings of a symposium, with contributions by the leading experts in their field.

Batra, S. W. T., and L. R. Batra. 1967. "The fungus gardens of insects." *Scientific American.* A general overview of the topic.

Benjamin, D. R. 1995. *Mushrooms: Poisons and Panaceas.* New York: W. H. Freeman. Appropriately termed "A handbook for naturalists, mycologists, and physicians," this book succeeds in providing de-tailed information on medicinal properties claimed for mushrooms. It is highly informative and authoritative.

Heim, R. 1977. *Termites et Champignons.* Paris: Editions Boubée. A comprehensive, rather technical description of the interactions of termites and their mushrooms is presented in this volume. In French.

Hobbs, C. 1995. *Medicinal Mushrooms,* 2d ed. Santa Cruz, CA: Botanica Press. This book describes the medicinal properties claimed for over 100 species of mushrooms. Scientific findings are presented in detail and supported by an extensive bibliography.

Lincoff, G., and D. H. Mitchel. 1977. *Toxic and Hallucinogenic Mush-room Poisoning.* New York: Van Nostrand Reinhold. A thorough compilation of known facts on poisonous mushrooms, their toxins, and the symptoms of the intoxications they cause.

Ott, J., and J. Bigwood. 1978. *Teonanacatl—Hallucinogenic Mushrooms of North America.* Seattle: Madrona Publishers. Presented here are extracts from a conference that included many of the authorities on the role of hallucinogenic mushrooms in human societies.

Pirozynski, K. A., and D. L. Hawksworth. 1988. *Coevolution of Fungi with Plants and Animals.* New York: Academic Press. Did fungi evolve along with their plant and animal hosts? The question is discussed in scientific papers by authorities in the field.

Rice, M. 1991. "Fine paper from mushrooms." *Mushroom: The Journal of Wild Mushrooming,* 34: 21–22. How to make paper from mushrooms.

Rice, M., and D. Beebe. 1980. *Mushrooms for Color.* Eureka, CA: Mad River Press. How to use mushrooms for dyeing textiles.

Riedlinger, T. J., ed. 1990. *The Sacred Mushroom Seeker: Essays for R. Gordon Wasson.* Portland, OR: Dioscorides Press. A compilation of articles on Gordon Wasson, the founder of modern studies on "ethnomycology," the role of mushrooms in human societies.

Wasson, R. G. 1968. *Soma—The Divine Mushroom of Immortality.* New York: Harcourt Brace Jovanovich. Here is the linguistic and other evidence on which the author based his postulation that the ancient Hindus used *Amanita muscaria* as the intoxicant represented by the god Soma. Available in paperback.

———— 1980. *The Wondrous Mushroom: Mycolatry in Mesoamerica.* New York: McGraw-Hill. The authoritative treatment of the use of psychotropic mushrooms in Mexico, with a description of seances with the *curandera* María Sabina.

Wheeler, W. M. 1973. *The Fungus-Growing Ants of North America.* [Reprint of 1907 original.] New York: Dover Publications (180 Varick Street, New York, NY 10014). The classic description of research into this fascinating biological subject.

Journals

McIlvainea. North American Mycological Association, 3556 Oakwood, Ann Arbor, MI 48104–5213. Published by the North American Mycological Association and offered as a benefit of membership, this journal includes general and technical articles of interest to amateur mycologists.

Mushroom: The Journal of Wild Mushrooming. P. O. Box 3156, University

Station, Moscow, ID 83843. A quarterly magazine for amateur mycologists, presenting articles on collecting, preserving, and cooking wild mushrooms.

Mycologist. Cambridge University Press, Cambridge, England. This full-color quarterly international journal published by the British Mycological Society provides articles on science, industry, and conservation aimed at a general audience.

Electronic Resources

Several services for mycologists are available on the Internet, including bulletin boards, mailing lists on general and specific topics, and databases. Some mushroom clubs are making their newsletters accessible electronically. Because these resources are changing rapidly, any addresses published here would be obsolete before very long. Searching the net under "mycology" will reveal most of the currently available services. More guidance may be available from your local or regional mushroom club.

Forays and Other Events

North American Mycological Association. This is the only amateur association that operates on a national level in North America. The association publishes a newsletter and a magazine. Contact: North American Mycological Association, 3556 Oakwood, Ann Arbor, MI 48104-5213; e-mail: <kwcee@umich.edu>.

National Forays. Every year the North American Mycological Association sponsors a foray that is attended by nearly 300 people, including professional mycologists. A different location is chosen for each trip.

Regional Forays. Every August, a three-day foray is held in the Northeast under the sponsorship of one of fourteen clubs in the region. About 250 people are usually in attendance, again including professional mycologists. For information, contact any of the northeastern clubs listed below. Other regional forays are sponsored by individual clubs or by regional organizations.

Conferences. The Telluride (Colorado) Mushroom Conference holds yearly meetings designed to expand people's knowledge of edible, poisonous, and psychoactive mushrooms. Write to Fungophile, Inc., P. O. Box 480503, Denver, CO 80248-0503.

North American Truffling Society. This is the only amateur association in North America that is dedicated to the pursuit and study of this specific kind of fungi. For information, write: North American Truffling Society, P. O. Box 296, Corvallis, OR 97339-0296.

Mushroom Clubs

Regional clubs sponsor mushroom walks, usually in their own vicinity. The frequency of walks and indoor meetings devoted to subjects of interest to amateur mycologists varies with each club. Some clubs sponsor "Fungus Fairs," exhibits of wild mushrooms for the general public. All clubs publish newsletters or bulletins. Newcomers are universally welcome.

Alaska

Alaska Mycological Society
P. O. Box 2526
Homer, AK 99603-2526

Glacier Bay Mycological Society
P. O. Box 65
Gustavus, AK 99826-0065

Southeastern Alaska
Mycological Society
P. O. Box 956
Sitka, AK 99835

Arizona

Arizona Mushroom Club
3060 East Minton Street
Mesa, AZ 85213-1698

Arkansas

Arkansas Mycological Society
5115 South Main Street
Pine Bluff, AR 71601-7452

California

Fungus Federation of Santa
Cruz
1305 East Cliff Drive
Santa Cruz, CA 95062-3722

Humboldt Bay Mycological
Society
P. O. Box 4419
Arcata, CA 95521-1419

Los Angeles Mycological Society
4907 South Maymont Drive
Los Angeles, CA 90032-4221

Mount Shasta Mycological
Society
623 Pony Trail
Mount Shasta, CA 96067-9769

Mycological Society of San
Francisco
P. O. Box 882163
San Francisco, CA 94188

San Diego Mycological Society
4215 Everts Street
San Diego, CA 92109

Sonoma County Mycological
Association
406 Pleasant Hill Road
Sebastopol, CA 95472

Colorado

Colorado Mycological Society
P. O. Box 9621
Denver, CO 80209-0621

Pikes Peak Mycological Society
1885 Ponder Heights Drive
Colorado Springs, CO 80906

Connecticut

Connecticut Mycological
Association
4 Glenn Place
Hastings-on-Hudson, NY 10706

Connecticut Valley Mycological
Society
59 Quarry Dock Road
Trumbull, CT 06611

Nutmeg Mycological Society
7 Stone Hill Drive
Westerly, RI 02891

District of Columbia

Mycological Association of
Washington
R. R. 4, Box 352
Luray, VA 22835-9037

Idaho

North Idaho Mycological
Association
7035 Cougar Gulch Road
Coeur D'Alene, ID 83814

Palouse Mycological Association
904 North Cleveland Street
Moscow, ID 83843-9407

Southern Idaho Mycological
Association
P. O. Box 843
Boise, ID 83701-0843

Illinois

Illinois Mycological Association
831 Renaissance Drive
Carol Stream, IL 60188-4437

Iowa

Prairie States Mushroom Club
310 Central Drive
Pella, IA 50219-1901

Kansas

Kaw Valley Mycological Society
2431 Atchison Avenue
Lawrence, KS 66047

Louisiana

Gulf States Mycological Society
414 Kent Avenue
Metaire, LA 70001-4324

Maine

Maine Mycological Society
R. R. 2, Box 3920
Bowdoinham, ME 04008-9619

Maryland

Lower East Shore Mushroom
Club
14011 Cooley Road
Princess Ann, MD 21853-3247

Massachusetts

Berkshire Mycological Society
Pleasant Valley Sanctuary
Lenox, MA 02140

Boston Mycological Club
6 Oak Ridge Dr. #4
Maynard, MA 01754

Michigan

Albion Mushroom Club
Whitehouse Nature Center
Albion, MI 49224

Lewiston Fun Country MHC
P. O. Box 783
Lewiston, MI 49756-0783

Michigan Mushroom Hunters
Club
2297 19th Street
Wyandotte, MI 48192-4125

West Michigan Mycological
Society
923 East Ludington Avenue
Ludington, MI 49431-2437

Minnesota

Minnesota Mycological Society
4424 Judson Lane
Edina, MN 55435-1621

Missouri

Missouri Mycological Society
2888 Ossenfort Road
Glencoe, MO 63038-1716

Montana

Southwest Montana
Mycological Association
3349 Trail Creek Road
Bozeman, MT 59715-8045

Western Montana Mycological
Association
P. O. Box 7306
Missoula, MT 59801

New Hampshire

Monadnock Mushroomers
Unlimited
P. O. Box 192
Alstead, NH 03602

Montshire Mycological Club
P. O. Box 59
Sunapee, NH 03782-0059

New Hampshire Mycological
Society
88 Cannongate III
Nashua, NH 03036-1948

New Jersey

New Jersey Mycological
Association
19 Oak Avenue
Denville, NJ 07834-2223

New Mexico

New Mexico Mycological
Society
1511 Marble Avenue N.W.
Albuquerque, NM 87104-1347

New York

Central New York Mycological
Society
343 Randolph Street
Syracuse, NY 13205-2357

Long Island Mycological Club
128 Audley Street
Kew Gardens, NY 11418

Mid Hudson Mycological
Association
1846 Route 32
Modena, NY 12548

Mid York Mycological Society
P. O. Box 164
Clinton, NY 13323-0164

New York Mycological Society
140 West 13th Street
New York, NY 10011-7802

Rochester Area Mycological
Association
54 Roosevelt Road
Rochester, NY 14618

North Carolina

Asheville Mushroom Club
51 Kentwood Lane
Pisgah Forest, NC 28768-9511

Blue Ridge Mushroom Club
P. O. Box 1107
North Wilkesboro, NC
48659-1107

Cape Fear Mycological Society
6132 Rock Creek Road N.E.
Leland, NC 28451

Triangle Area Mushroom Club
P. O. Box 61061
Durham, NC 27715-1061

North Dakota

Fun Gi's Mycological
Association
P. O. Box 2436
Bismarck, ND 58502-2436

Ohio

Ohio Mushroom Society
10489 Barchester
Concord, OH 44077

Oregon

Lincoln County Mycological
Society
6504 Southwest Inlet Avenue
Lincoln City, OR 97367-1140

Mount Mazama Mycological
Association
417 Garfield Street
Medford, OR 97501-4465

Oregon Coast Mycological
Society
P. O. Box 1590
Florence, OR 97439-0103

Oregon Mycological Society
13716 Southeast Oatfield Road
Milwaukie, OR 97222-7035

Pacific Northwest Key Council
1943 Southeast Locust Avenue
Portland, OR 97214-4826

Willamette Valley Mushroom
Society
1454 Manzanita Street S.E.
Salem, OR 97303-1915

Pennsylvania

Eastern Pennsylvania
Mushroomers
3119 Parker Drive
Lancaster, PA 17601-1634

Susquehanna Valley
Mycological Society
R. R. 1, Box 501
Brackney, PA 18812-8801

South Carolina

South Carolina Exotic
Mushroom Hunters
263 Long Pine Court
Chapin, SC 29063

Texas

Texas Mycological Society
7445 Dillon Street
Houston, TX 77061-2721

Utah

Mushroom Society of Utah
600 South University Street
Salt Lake City, UT 84102

Vermont

Vermont Mycology Club
P. O. Box 792
Burlington, VT 05402-0792

Washington

Jefferson County Mycological
Society
60 Tala Shores Drive
Port Ludlow, WA 98365

Kitsap Peninsula Mycological
Society
P. O. Box 265
Bremerton, WA 98310-0054

Northwest Mushroomers
Association
4320 Dumas Avenue
Bellingham, WA 98226-3007

Olympic Mountain Mycological
Society
P. O. Box 720
Forks, WA 98331-0720

Puget Sound Mycological
Society
University of Washington
Urban Horticulture GF-15
Seattle, WA 98195-0001

Snohomish County Mycological
Society
P. O. Box 2822
Everett, WA 98203-9822

South Sound Mushroom Club
6439 32nd Avenue N.W.
Olympia, WA 98502-9519

Spokane Mushroom Club
P. O. Box 2791
Spokane, WA 99220-2791

Tacoma Mushroom Society
P. O. Box 99577
Tacoma, WA 98499-0577

Tri-Cities Mycological Society
Route 1, Box 525C
Richland, WA 99352

Wenatchee Valley Mushroom
Society
P. O. Box 296
Monitor, WA 98836-0296

Wisconsin

Northwestern Wisconsin
Mycological Society
R. R. 3, Box 17
Frederic, WI 54837-9700

Parkside Mycological Club
5219 85th Street
Kenosha, WI 53142-2209

Wisconsin Mycological Society
5219 West Wells Street, Room
614
Milwaukee, WI 53233-1404

Canada

Cercle des Mycologues de
Montreal
4101 Rue Sherbrooke ESR, 125
Montreal, Quebec
Canada H1X 2B2

Cercle des Mycologues de
Rimouski
Université de Quebec, Rimouski
Rimouski, Quebec
Canada

Cercle des Mycologues de
Quebec
2000 Boulevard Montmorency
Quebec, Quebec
Canada G1J 5E7

Cercle des Mycologues du
Saguenay
CP 821
Chicoutimi, Quebec
Canada G7H 5EB

Chibougamau Mycological Club
804 5me Rue
Chibougamau, Quebec
Canada G8P 1V4

Edmonton Mycological Club
7416 182nd Street
Edmonton, Alta
Canada T5T 267

Mycological Society of Toronto
632 Durham Road 21, R. R. #4
Uxbridge, Ontario
Canada L9P 1R4

South Vancouver Island
Mycological Society
277 Durrance Road
Victoria, British Columbia
Canada V8X 4M6

Vancouver Mycological Society
17389 96th Avenue
Surrey, British Columbia
Canada V3S 5X7

CREDITS

Figures 1.1, 2.1, 2.2, 3.2, 3.3, 3.4, 3.5, 8.3, 9.1, 9.3, and 9.4 were created by John D. Woolsey.

Figure 3.1 was adapted by John D. Woolsey from C. T. Ingold, "Fan vaulting and agarics," *The Mycologist* 7 (1993): 58–59.

Figure 8.1 is reprinted from W. Robinson, *Mushroom Culture* (Philadelphia: McKay, 1870, p. 58).

Figure 9.3 was adapted by John D. Woolsey from C. T. Ingold, *The Fungal Spore* (Oxford: Clarendon Press, 1971).

The woodcuts of fungi which appear on pages iii, vii, ix, xi, 1, 3, 23, 41, 55, 57, 85, 97, 121, 123, 145, 161, 179, 181, 201, 220, 247, 255, 273, and 275 are from *Theatrum Fungorum oft Het Tooneel der Campernoelien* by Franciscus von Sterbeeck (Antwerp: Joseph Jacobs, 1675). Permission has been granted by the Farlow Reference Library of Cryptogamic Botany, Harvard University, Cambridge, Massachusetts for their use.

The drawing on p. 233 is from *The Insect Societies*, by E. O. Wilson (Cambridge, Mass.: Harvard University Press, 1971, p. 43). It depicts the inflated tips ("gongylidia") of the fungus grown and eaten by fungus-gardening ants. Each gongylidia is 30–50 μ in diameter.

273

The translations of haiku on pages 1, 55, 121, and 179 are by R. H. Blyth and are reprinted from *Transactions of the Asiatic Society of Japan*, 3rd series (1973) 2: 93–106.

The author wishes to thank Kenneth Kleene and Charles Hrbek for permission to reproduce their color slides.

Gills: *Cortinarius sp., Laccaria laccata, Daedalea quercina*, Elio Schaechter; *Collybia maculata*, Kenneth Kleene

Alternatives to Gills: *Crucibulum vulgare, Clavaria fusiformis, Peziza badioconfusa*, Charles Hrbek; *Geastrum saccatum*, Elio Schaechter

Color: *Chlorociboria aeruginascens*, Charles Hrbek; *Laccaria amethystina, Helotium citrinum*, Elio Schaechter; *Hygrocybe cantharellus*, Kenneth Kleene

Poisonous and Hallucinogenic Mushrooms: *Gymnopilus spectabilis, Amanita virosa*, Elio Schaechter; *Amanita muscaria*, Charles Hrbek

Edible Mushrooms: *Leccinum aurantiacum, Hypomyces lactifluorum, Craterellus fallax*, Charles Hrbek; *Cantharellus cibarius* (group), *Morchella esculenta*, Elio Schaechter; *Cantharellus cibarius* (detail), Kenneth Kleene

Bracket Fungi: *Polyporus squamosus, Pycnoporus cinnabarinus, Trametes versicolor*, Charles Hrbek; *Fomes fomentarius, Trametes versicolor* (detail), Elio Schaechter

INDEX

African-Americans, 218–219
agarick, 197, 198
Agaricus arvensis, 18, 65
Agaricus augustus, 65
Agaricus bisporus, 123, 147
Agaricus campestris, 65, 88, 126, 215
Agaricus crocodilinus, 89
Agaricus species, 88–89
Aleuria aurantia, 65
Alice in Wonderland, 37, 193
Allegro, John, 188
allergy to mushrooms, 215
amadou, 228–229
Amanita caesarea, 7, 210
Amanita calyptrata, 64, 137, 210
Amanita citrina, 76
Amanita flavoconia, 75
Amanita hemibapha, 137
Amanita muscaria, 7, 64, 75, 190–191, 194, 205, 207, 216
Amanita ocreata, 64, 210

Amanita pantherina, 216
Amanita phalloides, 7, 64, 76, 204, 210, 216
Amanita princeps, 207
Amanita rubescens, 64, 76
Amanita virosa, 64, 67, 207, 216
Amanita species, 64, 74, 209, 211
amanitins, 209, 212, 213
Anna Karenina (Tolstoy), 80–81
ants, leaf-cutting, 237–242
Argentina, 175
Armillaria bulbosa, 30
Armillaria mellea, 64, 84, 126, 134–135, 214
Armillaria ponderosa, 65
Arora, David, 86, 169
artist's fungus. See *Ganoderma applanatum*
Ascomycetes, 39, 166–167
ascus, 166
Aspergillus, 12, 167
Astraeus hygrometricus, 51

276   *Index*

Auricularia polytricha, 136–137,
     151, 196
Austria, 193

basidium, 44, 167
beefsteak fungus. See *Fistulina
     hepatica*
beetles, ambrosia, 242–243
Bible, 4
bird's-nest fungi, 52–54, 64
blewit, 18, 64
boletes, 38, 58, 80–82, 138, 227
Boletus edulis, 59, 65, 126, 130, 131
Boletus frostii, 82
Boston Mycological Club, 99, 114,
     173, 202
Boston Poison Center, 183, 217,
     219
bracket fungi. *See* polypores
brick caps. See *Naematoloma
     sublateritium*
Bridgeoporus nobilissimus, 227
British Isles, 14, 18, 112–115
Brodie, H. J., 52

Caesar's mushroom, 7, 64, 137, 210
Cage, John, 33, 111
Calvatia booniana, 137
Calvatia craniforme, 137, 139
Calvatia gigantea, 64, 86, 137
Calvatia sculpta, 137
Candida, 39, 167
Cantharellus cibarius, 64, 84, 126,
     132
chanterelles. See *Cantharellus
     cibarius*
Chestnut Blight, 26

chicken mushroom. See *Laetiporus
     sulphureus*
Chile, 128, 175–176
China, 128, 137, 148, 173, 196, 231
Chlorociboria aeruginascens, 225
Chlorophyllum molybdites, 89, 215,
     216
Ciccarelli, Alfonso, 8, 9
Claudius, murder of, 7, 204
Clitocybe nuda, 64, 126
Clusius, 9–11, 170
coccora. See *Amanita calyptrata*
Cooke, Mordecai, 113–115
Coprinus atramentarius, 91, 137, 218
Coprinus comatus, 64, 89, 137
Coprinus micaceus, 48, 90, 137
coral mushrooms, 39, 49, 83, 128
corn smut, 174
Cortinarius, for dyeing, 231
Craterellus cornucopioides, 64, 126
Craterellus fallax, 64, 84, 126
crimini, 147. See also *Agaricus
     bisporus*
Crucibulum species, 64
cryptogams, 29
Cyathus species, 64
cystidia, 48
Cyttaria darwinii, 175

Darwin, Charles, 175
death cup. See *Amanita phalloides*
*Dermocybe phoenicea var.
     occidentalis*, 231
desert truffle. See *terfez*
destroying angel, 64, 67, 210
Dictyophora species, 65
dikaryon, 35

DNA, fungal, 14, 26
dry rot, 223–224
Dutch elm disease, 26

earth stars, 51, 64
enoki. See *Flammulina velutipes*
ergot, 182, 195
Exidia glandulosa, 65
extremophiles, 248

fairy ring mushroom. See
  Marasmius oreades
false morel. See *Gyromitra esculenta*
Fistulina hepatica, 64
Flammulina velutipes, 64, 148, 152
flavors, mushroom, 123–125
fly agaric. See *Amanita muscaria*
Fomes fomentarius, 228, 229
Fossey, Dian, 21
fossil fungi, 25–26
France, 146, 162–164, 193
Fries, Elias, 13

Galerina, 214
Galerina autumnalis, 190
galls, plant, 243
Ganoderma applanatum, 21, 64,
  230
Ganoderma lucidum, 230
Gasteromycetes, 50
Geastrum fornicatum, 51
genes, fungal, 32–33
Gerard's Herbal, 11, 170, 197
Germany, 15, 130
Greek, ancient, 6, 29, 187, 197, 203
Grifola frondosa, 64, 126, 135–136,
  157, 158, 231

Guitry, Sacha, 205
Gymnopilus spectabilis, 183–186
Gyromitra esculenta, 217

hallucinogenic mushrooms, 183–195
Hebeloma syriense, 206
Hebeloma vinosophyllum, 205
hedgehog mushroom. See *Hydnum
  repandum*
Helvella lacunosa, 137
hen-of-the-woods. See *Grifola
  frondosa*
honey mushroom. See *Armillaria
  mellea*
horn-of-plenty mushrooms. See
  Craterellus cornucopioides,
  Craterellus fallax
huitlacoche. *See* corn smut
Hungary, 193
Hydnum coralloides, 137
Hydnum erinaceum, 137
Hydnum repandum, 64, 137
hydnums, 38
Hygrocybe coccinea, 84
Hygrocybe psittacina, 84
Hygrophorus fuligineus, 140–141
Hygrophorus goetzii, 247
Hypomyces lactifluorum, 64, 137

ibotenic acid, 191, 216
inky caps. See *Coprinus*
Inuit, 18
Italy, 161–164, 208

Jack O'Lantern mushroom. See
  *Omphalotus illudens, Omphalotus
  olearius, Omphalotus olivascens*

Japan, 20, 21, 105–106, 153, 156–160, 231
jelly fungi, 39

Kennett Square, Penn., 146
king bolete. See *Boletus edulis*
Kombucha, 198
Koryaks, 191

Laccaria trullisata, 63
Lactarius deliciosus, 58–59, 77, 79
Lactarius indigo, xii, 78
Lactarius lygniotus, 78
Lactarius species, 137
Laetiporus sulphureus, 64, 126, 135, 203, 232
Laughing Jim. See *Gymnopilus spectabilis*
LBMs ("little brown mushrooms"), 60–61, 91
Leccinum scabrum, 81
Lentinus edodes, 65, 127, 147, 157, 196, 222
Lentinus lepideus, 222
Lepiota americana, 126
Lepiota josserandii, 214
Lepiota procera, 126, 214
Lepiota rhacodes, 89, 126
Lepista nuda, 18
Lewis, Margaret, 98–99, 101, 104, 134
light, emitted by fungi, 16–17
Lincoff, Gary, 30, 111, 158, 207
Linnaeus, 58, 199
lobster mushroom. See *Hypomyces lactifluorum*

Lycoperdon, 87
Lycoperdon perlatum, 137
Lyophyllum decastes, 157

Marasmius oreades, 64, 92–93
matsutake, 21, 65, 105–106, 126, 127, 153
McIlvaine, Charles, 221
meadow mushroom. See *Agaricus campestris*
Mexico, 6, 174, 188–189, 230
Micheli, Pietro Antonio, 11–13
milk caps. See *Lactarius*
Millman, Larry, 18, 68
Morchella esculenta, 64, 126
morel cultivation, 149
morels, 62, 126, 129–131, 148, 166
muscarine, 216
muscimol, 191, 216
mushroom development, 28–32, 36–37, 40
mushroom picking, commercial, 154–156
mushroom stone, 177–178
mushrooms in coats of arms, 9
Mutinus caninus, 64
Mutinus elegans, 64
mycelium, 30
Mycena corticola, 60
mycology, 24
mycophobia, 14, 18
mycorrhiza, 163, 225–226
Mycota, fungal kingdom, 24

Naematoloma fasciculare, 65
Naematoloma sublateritium, 64

Native Americans, 17, 19, 192, 193, 198–199, 229
New Guinea, 192

odors, mushroom, 67–68, 153, 168
Old Man of the Woods, 65
Omphalotus olearius, 16
Omphalotus olivascens, 16
Omphalotus illudens, 65
orange peel fungus. See *Aleuria aurantia*
Oregon white truffle, 126, 165
oyster mushroom. See *Pleurotus ostreatus*

paddy straw mushroom. See *Volvariella volvacea*
parasol mushroom. See *Lepiota*
Paxillus involutus, 127
Phaeolus schweinitzii, 231
Phallales, 168
Phallus hadriani, 8
Phallus impudicus, 170, 172
Phallus species, 64. *See also* stinkhorn
Pholiota glutinosa, 157
pietra fungaia. See mushroom stone
pigs, truffling. See truffle pigs
Piptoporus betulinum, 229
Pleurotus ostreatus, 64, 95, 126, 127, 133–134, 147
Pliny the Elder, 6, 8, 169, 197, 203
Poland, 106
polypores, 227–232
Polyporus officinalis, 197
Polyporus tuberaster, 177

Pompei mushroom fresco, 59
porcini. See *Boletus edulis*
portobello, 146–147, 148. See also *Agaricus bisporus*
prince mushroom. See *Agaricus augustus*
Psilocybe, 189, 214, 216
psilocybin, 216
puffballs, 39, 50, 64, 86–88, 137, 199

recycling by fungi, 27–28
reishi. See *Ganoderma lucidum*
Romans, ancient, 6, 17, 29, 58, 197, 203
Russia, 19–20, 127, 183, 190–192, 193
Russula rubra, 128
Russula species, 76

Sabina, Maria, 187
saddle fungi, 137
Sayers, Dorothy, 205
Scandinavia, 17, 199
Schizophyllum commune, 31
Schultes, Richard, 187
Scleroderma citrinum, 64, 87
sclerotium, 176
Septobasidium, 244
Serpula lacrymans, 223
sex in the fungi, 30, 31–33, 35
shaggy mane. See *Coprinus comatus*
shiitake, 65, 141–142, 147, 150, 157, 158, 196, 222
Siberia, 190–192
slippery jack, 62, 65

Smith, Alexander, 66, 100, 178
Southeast Asia, 20, 207, 208
Spain, 57–59, 206
Sparassis radicata, 137
Sparassis crispa, 64, 137
spore dispersal, 42–54
spore print, 74
Steinpilz, 81
stinkhorn, 65, 168–173
Stone Age man, 3, 229
Strobilomyces floccopus, 65
Stropharia rugoso-annulata, 65, 91, 150
Suillus luteus, 62, 65

Talmud, 5
terfez, 165–166
Terfezia leonis, 165
termites, 233–237
Termitomyces titanicus, 237
Termitomyces, 235, 237
Thoreau, H. D., 41, 101
Trametes odora, 199
Trametes versicolor, 65, 228
Tremella mesenterica, 65
Tricholoma flavovirens, 137
Tricholoma magnivelare, 126, 153
Tricholoma matsutake, 153
Tricholomopsis platyphylla, 217
truffle hounds, 162, 164

truffle pigs, 161, 162
truffles, 161–168; cultivation, 163-164; desert, 165–166
Tuber gibbosum, 165
Tuber magnatum, 161
Tuber melanosporum, 163
tuckahoe, 176
turkey tail. See *Trametes versicolor*
Tylopilus felleus, 82

Ustilago maydis, 174

Volvariella volvacea, 148
von Schweinitz, L., 102–103

Wasson, Gordon, 187–189
waxy caps. See *Hygrophorus*, *Hygrocybe*
white button mushroom, 88, 145, 150. See also *Agaricus bisporus*
wine cap mushroom. See *Stropharia rugoso-annulata*
witch's butter. See *Tremella mesenterica*
Wolfiporia cocos, 177
wood ear mushroom. See *Auricularia polytricha*
Woolhope Naturalists Club, 112–114